AUTOMOTIVE BRAKE SYSTEMS

BOSCH

Impressum

Published by:
© Robert Bosch GmbH, 1995
Postfach 30 02 20
D-70442 Stuttgart
Automotive Equipment Business Sector,
Department for Technical Information (KH/VDT).
Management: Dipl.-Ing.(FH) Ulrich Adler.

Editor-in-Chief:
Dipl.-Ing. (FH) Horst Bauer.

Editors:
Dipl.-Ing. (FH) Anton Beer,
Ing. (grad.) Arne Cypra.

Layout:
Dipl.-Ing. (FH) Ulrich Adler,
Joachim Kaiser.

Technical graphics:
Bauer & Partner, Stuttgart.

Translation:
Peter Girling.
Translated by:
Ingenieurbüro für Technische und
Wissenschaftliche Übersetzungen
Dr. W.-D. Haehl GmbH, Stuttgart
Member of the ALPNET Services Group
William D. Lyon M.A.

Printed in Germany. Imprimé en Allemagne.
1st Edition, August 1995, based on
1st German Edition, October 1994.

Distribution:
SAE Society of Automotive Engineers
400 Commonwealth Drive
Warrendale, PA 15096-001 USA

ISBN 1-56091-708-3

Robert Bentley, Publishers
1000 Massachusetts Avenue
Cambridge, MA 02138
1-800-423-4595 USA

ISBN 0-8376-0331-5

Authors

Brake-system basics
Dipl.-Ing. E. Siegert, Mercedes-Benz AG, Stuttgart.
Dr.-Ing. E.-C. v. Glasner, Mercedes-Benz AG,
Stuttgart.
Dipl.-Ing. H. Geißler, Mercedes-Benz AG, Stuttgart.
Ing. (grad.) P. Berg in cooperation with the respon-
sible technical departments of Robert Bosch GmbH.

Brake systems for passenger cars
Dipl.-Ing. R. Becker, Dipl.-Ing.(FH) J. Pfäffle
(Brake systems).
Dipl.-Ing.(FH) P. Sowa (Brake-system components).
Dipl.-Ing. A. Czinczel, Dr.-Ing. G. Schmidt,
Dipl.-Ing. J. Gerstenmeier, Dipl.-Ing. A. Knust,
Dipl.-Ing. K. Kühner, Dipl.-Ing.(FH) K.-D. Reinke,
Dipl.-Ing. A. Stegmaier (Antilock Braking System
ABS).
Dipl.-Ing. M. Meißner, Dipl.-Ing. A. Sigl,
(Traction control ASR).

Commercial vehicles – Basic concepts, systems, and layouts
Ing. (grad.) P. Berg, in cooperation with
the responsible technical departments of the
Robert Bosch GmbH.

Compressed-air equipment – Symbols
Editorial processing in cooperation with the
responsible technical departments of the Robert
Bosch GmbH.

Equipment for commercial vehicles
Ing. (grad.) P. Berg in cooperation with the
responsible technical departments of the Robert
Bosch GmbH.

Brake testing
Inspection and maintenance procedures: Editorial
processing in cooperation with the responsible
technical departments of the Robert Bosch GmbH.
Brake test stands: Ing. (grad.) P. Berg, in
cooperation with the responsible technical departments of the Robert Bosch GmbH.

Unless otherwise stated, the above are all employ-
ees of Robert Bosch GmbH, Stuttgart.

Foreword

This reference book is compiled from the most important manuals in the publication series "Bosch Technical Instruction" insofar as they concern automotive brakes. The subject matter covered has been expanded to include sections dealing with basic concepts and with maintenance. This book is intended to satisfy the "thirst for knowledge" of a broad circle of technically interested readers.

With automotive brakes, safety is the prime consideration. State-of-the-art technology in the chassis and brake sectors, backed-up by electronic control engineering provides for reliable steerability and stability, not only when braking but also when driving off and when accelerating – provided that the physical limits are not exceeded.

Today's modern automobiles and commercial vehicles are equipped with reliable, high-performance brake systems which ensure efficient braking even at very high speeds. But even the best brakes cannot prevent a driver from hitting the brake pedal too hard when faced by severe road conditions or a panic situation.

Even under such circumstances, the antilock braking system (ABS) from Bosch ensures that the driver not only retains steerability, but also full control of the vehicle. Traction control (ASR) also reduces the strain on the driver by preventing drive-wheel spin when driving off and accelerating. Both of these safety systems contribute to keeping the vehicle stable and steerable.

This book provides comprehensive information on the present standard of engineering in the fields of drive-off, acceleration, and braking. The reader who is technically interested in automotive engineering is provided with a detailed but easily understood description of the most important brake-system components and of their method of functioning.

Summary

Brake-system basics

Terminology, design and structure

(based on ISO 611)

Purpose

Brake systems discharge the following functions:
- reducing vehicle speed,
- bringing a moving vehicle to a halt, and
- keeping a halted vehicle stationary.

This means that brake systems play a vital role in making motor vehicles suitable for practical application. They are essential for ensuring highway safety, which is why brake systems are subject to strict official regulations. A vehicle's approval for homologation and highway operation is contingent upon compliance with a number of national and international regulations.

These laws define terminology, describe basic concepts and prescribe the minimum requirements to be met by the components within the brake system. The terminology and regulations in the following text are based on the legal frame-work as it exists in Germany and within the European Union. The text omits references to specific regulations applicable in other countries.

Braking equipment

The braking equipment of a vehicle includes all of its brake systems, that is, all of the systems responsible for reducing the velocity of a moving vehicle, reducing its rate of acceleration, increasing its rate of deceleration, halting the vehicle, and preventing the vehicle from resuming movement once it is stationary.

Brake systems

Application

Legal regulations stipulate that the brake systems on heavy commercial vehicles will include such equipment as:

- service brakes,
- secondary brakes,
- a parking brake, and (in some cases)
- a continuous-operation brake, and
- an automatic brake system.

The service and parking-brake systems on motor vehicles are equipped with separate control and transmission (force-relay) mechanisms. The service brakes are generally applied with a foot pedal, while the mechanism for engaging and releasing the parking brake can be either hand or foot-operated.

The secondary-brake system frequently shares components with the service and/or parking brakes; one example is when one circuit in a dual-circuit service-brake system assumes the role of a secondary-brake system.

Continuous-operation brake systems, or retarders, serve as supplementary, wear-resistant brake systems designed to assume the loads to which the service brakes would otherwise be exposed under extreme conditions, such as those encountered on extended downhill stretches. Automatic brake systems are only relevant for vehicles with trailers.

Service-brake system

The driver uses the service brakes to decelerate the vehicle at a desired rate ("gradual" or "hard" braking) during the course of normal operation. The brakes can be employed to maintain a constant vehicle speed during descents, or to bring the vehicle to a halt.

Secondary-brake system

In the event of a service-brake malfunction, the secondary-brake system assumes the function of the conventional service brakes by allowing the driver to decelerate the vehicle at a selected rate ("gradual" or "hard" braking), to reduce its rate of acceleration, or increase its rate of deceleration, and to bring the vehicle to a halt.

Parking-brake system

The parking brake furnishes a mechanical means of keeping the vehicle stationary, even on inclined road surfaces and – especially important – in the absence of the driver. A typical example is the hand-brake.

Continuous-operation brake system (retarders, etc.)

The retarder system incorporates all of those elements that allow the driver to initiate reductions in a selected speed, or to avoid exceeding it, the usual field of application being on extended down-grades. These continuous or sustained-operation brakes (retarders) are frequently installed in heavy commercial vehicles.

Automatic brake system

The automatic brake system consists of all those elements which automatically apply braking force to a trailer's wheels in the event of intentional or accidental separation from the towing vehicle.

Antilock Braking System (ABS)

The antilock braking system is the aggregate of all those components within the service-brake system responsible for providing closed-loop control of wheel-slip rates at one or several wheels during braking manoeuvres.

Braking force at wheels with direct control is regulated with the assistance of an individual sensor for that wheel. In systems with indirect control, data from the sensor(s) at the other wheel(s) are employed.

Components

Application

The brake system consists of:
- an energy-supplying system,
- a control system,
- a transmission (force-transfer) system for transferring brake-force control commands and for activating the engine, continuous-operation and parking brakes,
- supplementary system(s) for braking trailer(s), and
- wheel brakes.

Each of these components plays a role in determining the level of the forces that will ultimately define how effectively the vehicle (and any connected trailer) is braked.

Different application ranges for divergent vehicle types mean that brake systems are necessarily exposed to a highly variegated range of demands. The inevitable result is a multiplicity of highly diversified brake systems, differing in both design and components as well as in their ultimate operating environments.

Energy-supplying system

The energy-supplying section of the brake system consists of those components used to supply, control – and, if required, to convert – the energy required for braking.

This section ends where the transmission (force-transfer) system begins. That is, at the point where the individual brake-system circuits, including circuits for any ancillary equipment that may be fitted, are either safeguarded from the energy-supplying system, or safeguarded from each other.

The energy source may be located outside the vehicle (air compressor with pneumatic trailer brakes), and can also be in the form of human muscular force.

Control system

The control (or actuation) system comprises those brake-system components responsible for initiating and regulating system braking. The control signal can be conveyed with mechanical, pneumatic, hydraulic or electrical devices, and external energy sources may also be employed. The control system is defined as starting at that point in the brake system to which the actuation force is directly applied.

It can be activated:
– directly, with the foot or hand,
– via indirect driver intervention, or without any direct driver action (trailers only),
– through variations in the pressure or electrical current in a relay line between tractor and trailer when one of the tractor's brake systems is applied, or in response to malfunction,
– in response to a vehicle's mass inertia, its weight, or one of its major components.

The control system ends at the point where the braking energy is distributed, or where a portion of the available energy is diverted to control the braking energy.

Transmission (force-transfer) system

The transmission system comprises the brake-system components responsible for relaying the energy from the control system. It starts where the control or energy-supplying system terminates. It ends at the part of the brake system where the forces that directly oppose the vehicle's tendency to move are generated. The transmission system can be mechanical, hydropneumatic (positive or vacuum pressure), electric or hybrid (e.g., hydromechanical).

Brake

The brake itself consists of those brake-system components in which the forces that directly oppose the vehicle's tendency to move are generated. Examples are the friction brake (disc or drum) and continuous-action brakes (hydrodynamic and electrodynamic retarders, engine brakes).

Supplementary tractor systems for trailer towing

This supplementary system consists of those parts of the tractor's brake system employed to supply energy and provide control for the brake system(s) in the trailer. The supplementary system comprises the components between the tractor's energy-supplying system and the coupling head (up to and including the supply line) as well as those parts located between the tractor's transmission system(s) and up to and including the brake-line coupling head (gladhand).

Energy types and applied media

Application

Depending upon the type of energy applied to the brake system, a distinction is drawn between the:
– muscular-energy brake system,
– power-assisted brake system,
– power-brake system,
– inertial (overrun) brake system, and the
– gravity brake system.

These types of brake system can also be employed in various combinations. One example is the power-assisted brake system, which differs from the power-brake system in deriving a portion of its energy from the force applied to the brake pedal.

The power-assisted and power-brake systems also differ in the types of media employed in the transmission (force-transfer) circuits. Pneumatic (compressed air) and hydraulic (brake fluid) systems predominate, and electrical energy is also employed in isolated applications.

Muscular-energy brake system

Brake system in which the energy required to generate braking force is supplied solely by the physical effort of the driver.

Power-assisted brake system

Brake system in which the energy necessary to generate the braking force is supplied by the physical effort of the driver working in conjunction with one or more energy-supplying devices.

Power-brake system
Brake system in which the energy required to generate braking force is supplied by one or more devices producing a force completely independent of the driver's physical effort.
Note: This definition does not include brake systems in which the driver can apply muscular force to generate effective braking energy in the event that all energy-supplying devices should fail.

Inertial (overrun) brake system
Braking system in which the energy necessary to produce braking force is generated as the trailer overruns and approaches the tractor (towing vehicle).

Gravity brake system
Brake system in which the energy required to generate braking force is produced when a component of the trailer (such as the drawbar) moves downward in response to gravity.

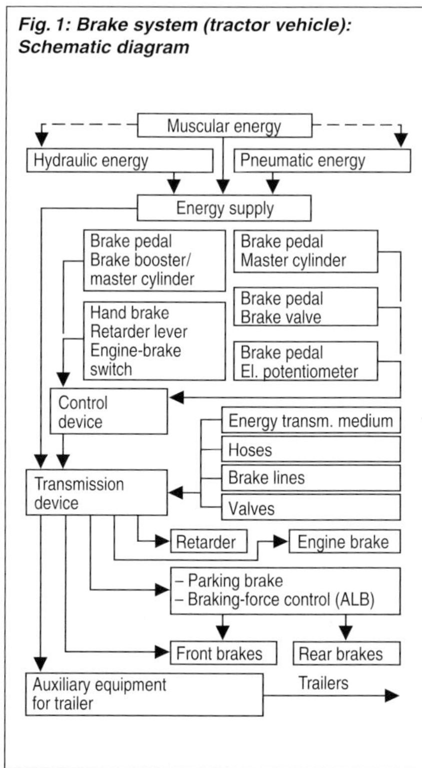

Fig. 1: Brake system (tractor vehicle): Schematic diagram

Transmission systems

Application
Mechanical, hydraulic, pneumatic and/or electrical/electronic methods are used to relay the required forces through the brake system. Various combinations can be used to transfer the forces to the wheel cylinder. Electrical and electronic transmission systems will play an increasingly significant role in the electropneumatic and electrohydraulic brake systems anticipated in the next vehicle generation. The transmission system can be of the single-circuit or multiple-circuit type.

Single-circuit brake system
In the single-circuit brake system, forces are transmitted along a single path; a failure in this circuit will thus make it impossible to relay the energy required for generating the braking power.

Multi-circuit brake system
The multi-circuit brake system embodies a design in which forces are transferred through two or more circuits; even in the event that one of the circuits fails, it remains possible to relay some or all of the energy required for generating the braking power.

Definitions relating to vehicle combinations

Single-line brake system
This is an assembly in which the brake systems of the individual vehicles are designed in such a way that a single dual-purpose line is employed alternately for supplying energy to, and for controlling the trailer's brake system.

Dual and multi-line brake systems
In this type of configuration, the brake systems of the individual vehicles are combined in such a way that the trailer brakes are controlled and supplied with energy by separate lines operating simultaneously.

Continuous braking system

This combination of the brake systems in tractor-trailer combinations incorporates the following features:
– with a single operation, and without moving from behind the wheel, the driver can directly operate a control device on the tractor and indirectly operate a control device in the trailer for modulated ("gradual" or "hard") braking;
– the energy used to brake the individual vehicles in the combination originates from a single energy source (this can be the muscular force exerted by the driver);
– simultaneous or suitably phased braking is available for each of the vehicles in the combination.

Semi-continuous brake system

This combination of the brake systems in tractor-trailer combinations incorporates the following:
– with a single operation, and without moving from behind the wheel, the driver can directly operate a control device on the tractor and indirectly operate a control device in the trailer for modulated ("gradual" or "hard") braking;
– the energy required to brake the individual vehicles in the combination is supplied by at least two different sources (one of which can be the muscular force exerted by the driver);
– simultaneous or suitably phased braking is available for each of the vehicles in the combination.

Non-continuous brake system

The non-continuous brake system is defined as a combination of brake systems in individual vehicles within a tractor-trailer combination in which neither continuous or semi-continuous principles apply.

Brake-system lines
(selection)

Wiring and conductors

These are used to transfer electrical energy.

Tubular lines

Pneumatic and hydraulic energy are conveyed through rigid, semi-rigid and flexible tubing.
Permanently formed <u>rigid tubing</u> provides connections between mutually stationary components. Any deformation occurring within such a tube will be permanent.
<u>Semi-rigid tubing</u> provides an elastic (deformable) coupling between two components in relative motion.
<u>Flexible lines</u> connect two components in a state of relative motion.

Supply line

The supply line is a special charging line (connecting the brake systems in the individual vehicles within a tractor-trailer combination). It transmits energy from the tractor to the energy-storage device(s) in the trailer.

Brake line

The brake line is a special control line (connecting the brake systems in the individual vehicles within a tractor-trailer combination). Through it, the energy required for control passes from the tractor to the energy-storage device(s) in the trailer.

Dual-function supply and brake line

This line combines the functions of the supply and brake lines (single-line brake system).

Secondary-brake line

Special working line transmitting the energy required for secondary trailer braking from the tractor to the trailer.

Basic physical concepts

Braking process

The standard defines braking as follows: "Processes occurring between initial actuation of the control system and the termination of the braking procedure."

Graduated braking
Braking which, within the normal operating range of the control device, continuously furnishes the driver with the option of initiating precisely variable increases and reductions in braking force by applying force to the control device.
If increased levels of application force at the control device increase the braking force, then lower levels of force shall lead to a reduction in braking force.

Brake-system hysteresis
Brake-system hysteresis is defined as the difference between the control forces used to apply and release the brakes at a constant braking torque.

Brake hysteresis
Brake hysteresis is defined as the difference in the application forces used to apply and release the wheel brake at constant braking torque.

Forces and torques

Control force F_c
This is the force level exerted at the control device.

Application force F_s
In friction-brake devices, the application force is the total force applied at the friction surface (pad or lining); the resulting friction generates the desired braking force.

Braking torque
The braking torque is the product of the frictional forces that result from the application force and the distance between the surfaces where these forces are effective and the axis of rotation.

Total braking force F_f
The total braking force F_f is the sum of the effective braking forces at all tire contact surfaces, as generated during operation of the brake system; these forces counteract the vehicle's movement or its tendency to move.

Braking-force distribution
The braking-force distribution provides a percentual definition of relative braking force exerted at each of the vehicle's individual axles as a component of total braking force F_f, e.g., 60 % front, 40 % rear.

Brake coefficient C^*
The brake coefficient C^* is the ratio of total circumferential force and the application force at the particular brake. Where the application forces vary between brake shoes, the mean should be employed for calculations.

Time factors

The following periods are illustrated in Figure 1.

Reaction time
The reaction time is the period that elapses between perception of the state or object requiring active response and active application of the system control device (t_0).

Actuating time of the control device
The actuating time of the control device is defined as the period elapsing between the moment at which force becomes effective at the control device (t_0) and the moment at which it reaches its final position corresponding to the applied control force or control travel (this applies mutatis mutandis for operations in which the brakes are released).

Initial response time $t_1 - t_0$

The initial response time defines the delay between application of force to the control mechanism and the onset of effective braking force (Fig. 1).

Pressure build-up time $t_1' - t_1$

This is the time that elapses between the moment when braking force becomes effective and the moment at which a defined value is obtained (in Fig. 1 this is 75 % of the asymptotic value of the wheel cylinder's internal pressure as defined in EU 71/320 EEC, Appendix III/2.4).

Composite response time

This is the sum of the initial response and build-up times. It is employed in evaluating brake-system response character- istics in the period leading up to effective application of maximum braking force.

Active braking time $t_4 - t_1$

The active braking time is the period that elapses between the time when braking force becomes effective and the moment at which it ceases. If the vehicle comes to a halt before termination of the brake application, the moment at which its movement ceases will define the end of the active braking time (Fig. 1).

Release time

The release time is the period that elapses between the initial force reduction at the control device and the termination of braking force.

Total braking time $t_4 - t_0$

The total braking time is the period that elapses between the initial application of force at the control device and termination of braking force at the brake. If the vehicle comes to a halt before termination of the brake application, the moment at which its movement ceases will define the end of the total braking time (Figure 1).

Braking distance s

The braking distance s is the distance travelled by the vehicle during the total braking time. If the end of the total braking time is defined by the point at which the vehicle halts, then the distance travelled up to this point is referred to as the "stopping distance."

Braking deceleration

The braking deceleration is the reduction in vehicle speed generated by the brake system within the time t. The "mean total deceleration" is the mean retardation in the period $t_3 - t_2$ within the fully developed deceleration process (Fig. 1).

Braking factor z

The braking factor is the ratio between the total braking force F_f and the static composite weight G_s being exerted on the vehicle's axles.

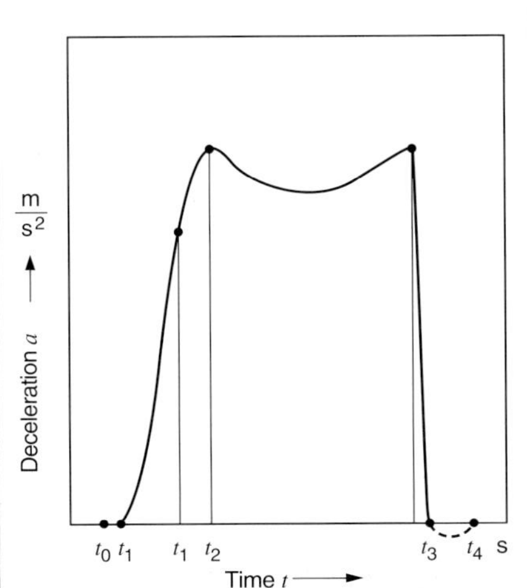

Fig. 1 Example for times and deceleration when braking to a stop

up to t_0	Reaction time
t_0:	Initial application of force to the control device
t_1	Start of deceleration
t_1'	Pressure build-up time completed
t_2	End of deceleration
t_3	Full deceleration ends
t_4	Braking completed (vehicle stationary)
$t_1 - t_0$	Initial response time
$t_1' - t_1$	Pressure build-up time
$t_3 - t_2$	"Mean fully developed deceleration" range
$t_4 - t_1$	Active braking time
$t_4 - t_0$	Total braking time

Legal regulations

General Certification or Type Approval (Allgemeine Betriebserlaubnis, or ABE)

In Germany, the official General Certification or type approval for vehicles is based on brake-system testing in accordance with examination procedures selected by the manufacturer who can choose testing based on:

– national regulations (in Germany, as defined in § 41 of the StVZO (FMVSS/CUR), together with the applicable guidelines for brake testing,

– directives of the European Union (EC Directive 71/320 EEC) and the applicable supplements and appendices, or

– ECE Directives 13 and 78 as issued by the UN Economic Commission in Geneva.

The requirements concerning application of antilock braking systems (ABS) defined in § 41b StVZO are more stringent than the corresponding EU regulations. § 41 of the StVZO, EC Directive 71/320 EEC and ECE Directive 13 are largely equivalent in all other areas.

Vehicle classifications

The categories used to define automotive brake systems are defined in EC Directive 71/320 EEC and ECE Directive 13.

Category L vehicles

Motor vehicles with less than four wheels, motorcycles (including mopeds and motor-driven cycles) and three-wheeled vehicles. Two and three-wheeled vehicles must be equipped with two mutually independent brake systems (Table 1).

Category M and N vehicles

Motor vehicles intended for transporting persons (M) or cargo (N), equipped with at least four wheels, or with three wheels and a gross vehicle weight of > 1 metric ton (Tables 2 and 3). Certain types of vehicle from category M_2 and N_2 upward must be equipped with ABS. Furthermore, to satisfy requirements for extended downhill driving, category M_3 and N_3 vehicles may also be equipped with continuous-action brake systems (mandatory on category M_3 buses of > 10 tons).

Category O trailers

Trailers and semi-trailers. Only a safety connection to the towing vehicle is required on Category O_1 (Table 4) trailers. Inertial brake systems are approved for Category O_2. Service and parking-brake systems are mandatory from category O_2 upward (shared components are allowed). ABS is required on certain types of trailers from category O_3 upward.

Vehicles with Antilock Braking System (ABS)

Where ABS is prescribed for vehicles of categories M_2, M_3 and N_2, N_3 compliance with category 1 as defined in EC Directive 71/320 EEC Appendix X is required. Only minimum requirements are defined for ABS-equipped trailers.

Table 1. Categorie L.

Grading	Type No. of wheels	Engine-displace-ment cm³	Maximum-speed km/h
L_1	2	≤ 50	≤ 50
L_2	3	≤ 50	≤ 50
L_3	2	> 50	> 50
L_4	3 [1]	> 50	> 50
L_5	3 [2]	> 50 [3]	> 50

[1] Asymmetrical to the vehicle's longitudinal axis.
[2] Symmetrical to the vehicle's longitudinal axis.
[3] ≤ 1 t gross vehicle weight.

Table 2. Category M.

Grading	Drivers seat + pass. seats	Gross veh. weight
M_1	1…9	
M_2	> 9	< 5 t
M_3	> 9	> 5 t

Table 3. Category N.

Grading	Gross veh. weight (t)
N_1	≤ 3.5 t
N_2	3.5…12 t
N_3	> 12 t

Table 4. Categorie O.

Grading	Gross veh. weight (t)
O_1 [1]	≤ 0.75 t
O_2	0.75…3.5 t
O_3	3.5…10 t
O_4	> 10 t

[1] Only for single-axle trailers.

Brake-circuit configurations

Variations

Legal regulations stipulate a dual-circuit transmission (force-transfer) system as mandatory equipment.

Although DIN 74000 describes five possible options, the II and X configurations have become standard, as relatively uncomplicated brake lines, hoses, connections, and static and dynamic sealing devices can be used to achieve reliability on a par with that furnished by a single-circuit brake system.

The potential for circuit failure arising from excess thermal loads at one of the wheel cylinders represents a potential liability of the HI, LL and HH configurations, as the failure of both wheel brakes at one wheel could lead to total brake failure.

Vehicles with a pronounced forward weight bias employ the X configuration to satisfy legal requirements for secondary-braking effects.

The II pattern is particularly well suited for use on rear-heavy vehicles as well as medium-heavy and heavy-duty commercial vehicles.

Configurations

II variant

Separate circuits for the front and rear axles – one circuit operates on the front wheels, and the other on the rear (Fig. 1a).

X variant

Diagonal distribution pattern. Each of the circuits operates at one front wheel and at the diagonally opposed rear wheel (Fig. 1b).

HI variant

One circuit for the front, and a second for the front and rear wheels. One brake circuit operates operates on both axles, while the other functions at the front wheels only (Fig. 1c).

LL variant

Front and rear/front and rear distribution pattern. Each of the brake circuits on both front and one rear wheel (Fig. 1 d).

HH variant

Front and rear/front and rear distribution pattern. Each of the circuits operates on both the front and the rear wheels (Fig. 1e).

Fig. 1: Braking-force distribution: Variants
a) II variant, b) X variant, c) HI variant,
d) LL variant, HH variant.
1 Brake circuit 1, 2 Brake circuit 2.
← Direction of travel.

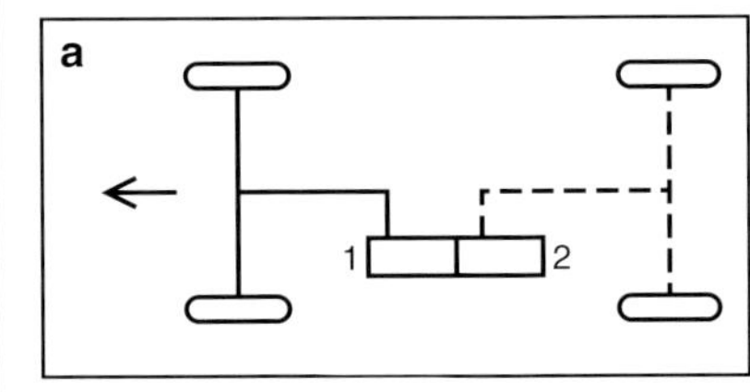

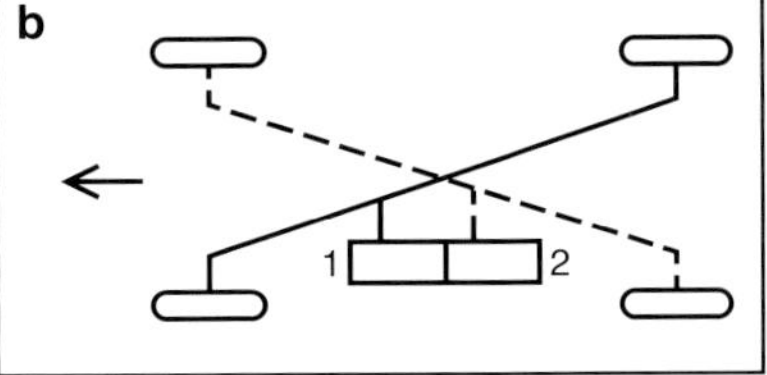

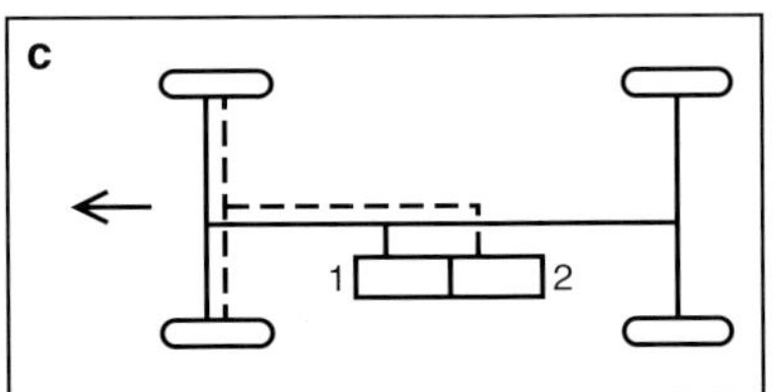

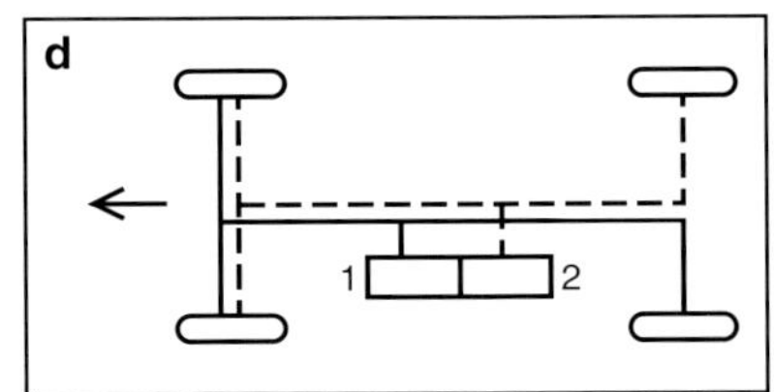

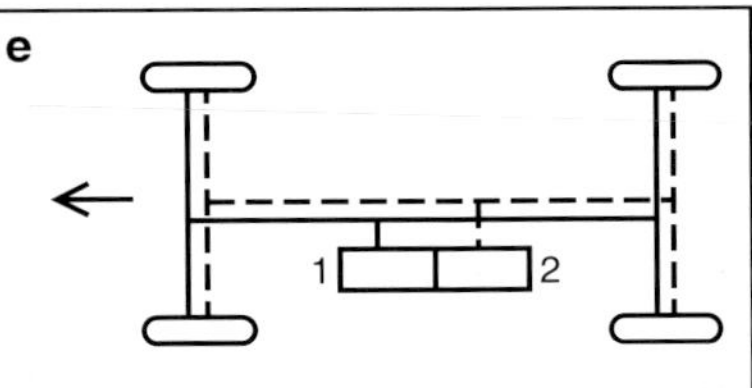

Operating dynamics

Kinetic processes

The persons and cargo transported within a vehicle remain in a (basically) motionless state relative to each other. While underway, however, both they and the vehicle are in a state of motion relative to the road which provides the fixed point of orientation.

It is possible to divide vehicular motion into two categories. These are constant (at a uniform velocity) and inconstant motion (drive-off/acceleration and braking/deceleration, with the associated variations in speed)

The kinetic energy required to propel the vehicle is generated by its engine.

To change the magnitude and direction of the vehicle's kinetic state, extraneous forces must be applied to the vehicle and/or reactive forces generated at the tire contact patch must be transferred to the road surface.

The effectiveness of the brakes depends upon the adhesion (coefficient of friction) between tires and road.

Synergetic complex: "Driver-Vehicle-Environment"

Regarding the dynamics of vehicular operation, operating reponse is generally defined as the cumulative response pattern of the composite system: "driver-vehicle-environment" (Fig. 1).

The driver, as the first element in this complex, employs the sum total of all subjective impressions as the basis for evaluating the quality of the vehicle's dynamic response.

In contrast, data derived from selected operating manoeuvres with no driver input (open-loop operation) provide an objective description of the vehicle's dynamic response patterns. This type of procedure entails replacing the driver – for which precise response definitions are still impossible – with processes designed to apply interference factors to the vehicle. The resulting vehicle reac-tion patterns are suitable for objective analysis and evaluation.

The following operating manoeuvres are based on definitive current or prospective ISO standards. These procedures (carried out on dry road surfaces) serve as recognized processes for evaluating vehicle stability:
– steady-state cornering (skidpad),
– transient response,
– braking during cornering,
– crosswind sensitivity,
– linear tracking and
– throttle-shift reactions on the skidpad.

To date, it has still not been possible to obtain comprehensive objective definitions for the dynamic characteristics encountered in closed-loop operation (Fig. 1), as adequate data on the precise control characteristics of the human element remain unavailable.

Steady-state cornering (skidpad)

The most important data to be derived from steady-state skidpad testing are the maximum rate of lateral acceleration, and the dynamic changes that accompany the variations in lateral acceleration in the range below the maximum rate. The vehicle's intrinsic handling response can be classified according to the following:
– understeer,
– oversteer, and
– neutral steering.

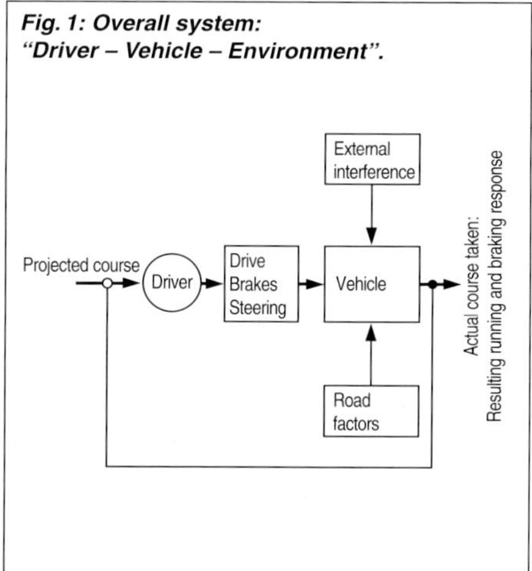

Fig. 1: Overall system: "Driver – Vehicle – Environment".

Transient response

Both the vehicle's steady-state handling, as well as its response characteristics during directional transitions are significant factors. These characteristics are relevant, for instance, when the driver of a vehicle traveling in a straight line must suddenly initiate evasive manoeuvres.

Braking during cornering

Of all the manoeuvres encountered in everyday driving, one of the most critical – and thus most significant for vehicle design – is braking during cornering. The vehicle's response pattern must represent the optimum compromise between steering response, stability, and brake effectiveness.

Parameters

The main criteria employed in evaluating vehicle dynamics (Fig. 2) are:
- steering-wheel angle,
- lateral acceleration,
- linear acceleration/deceleration,
- yaw velocity,
- float and roll angles.

Additional data are employed to verify and confirm the other available information on specific points of vehicle performance:
- linear and lateral velocity,
- steering angles at front and rear wheels,
- slip angles at all wheels,
- force applied at steering wheel.

Dynamic response of heavy commercial vehicles

Objective evaluations of a commerical vehicle's dynamic response characteristics are based on various driving manoeuvres including steady-state skid-pad testing, response to sudden sharp changes in steering angle (reaction when wheel is turned sharply at a defined angle) and braking during cornering.

The dynamic lateral response of truck and trailer combinations generally differs from that of single vehicles. Particularly significant are the load distribution between tractor and trailer, and the design and geometry of the mechanical coupling used to join the combination.

The most critical case is represented by a combination in which an unladen truck tows a laden center-axle trailer. This type of combination demands a high degree of driver skill.

On tractor-semitrailer combinations, braking manoeuvres initiated under extreme conditions can lead to jackknifing. This process starts as a loss of lateral traction at the tractor's rear axle; this can occur in response to excessive braking on slippery road surfaces or due to excess yaw on "μ-split" surfaces (e.g., varying traction coefficients at the crown of the road and at the edge of the lane). An anti-lock braking system can be installed to inhibit jackknifing.

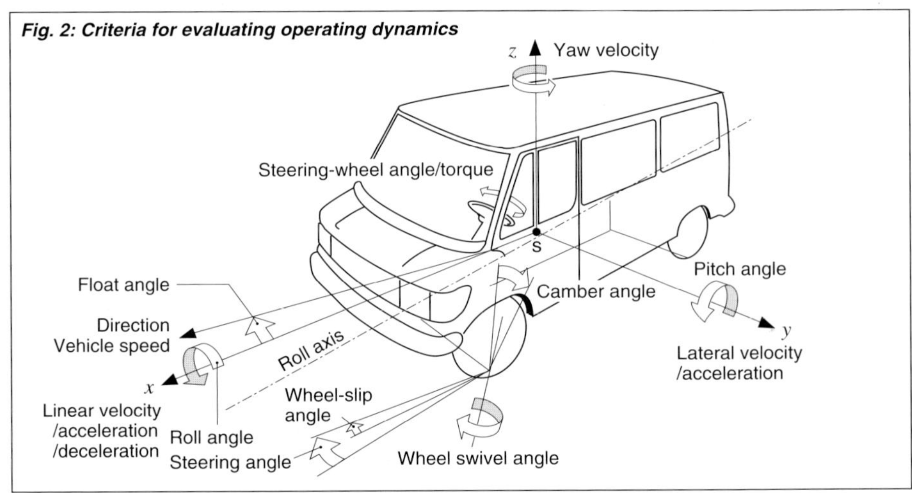

Fig. 2: Criteria for evaluating operating dynamics

Tires and road surface

Purpose
The tire, as the connecting element between vehicle and road surface, is a vital factor in vehicle safety.

The tire transfers tractive, braking and lateral forces during processes in which physical factors determine the actual limits of a vehicle's ability to cope with dynamic loads.

The salient characteristics used to evaluate tires are:
- straight-running stability,
- cornering stability,
- traction on road surfaces affording various levels of adhesion,
- traction in various weather conditions,
- steering performance,
- comfort (suspension, noise),
- durability, and
- economic factors.

Tire structure
The extensive range of tire designs can be classified according to development level and the technology which they incorporate. Designs are determined with reference to the various operating and emergency characteristics required of a conventional vehicle tire. As one example, the cords in the carcass(es) of a radial tire – which has advanced to become the standard in passenger-car applications – run in the shortest and most direct path – radially – from bead to bead. A belt surrounds this relatively thin, flexible carcass to provide the required stability.

On tubeless tires, the tube is replaced by a vulcanized, airtight inner liner.

Tire specifications
The essential performance data – such as tire dimensions, load capacity, required inflation pressures, speed rating – have been standardized to facilitate interchangeability (Table 1).

Legal requirements
Official regulations and guidelines define the tire specifications and determine which tires shall be used under any given conditions. The standards specify the maximum approved operating speeds and define tire classifications.

According to § 36 of the StVZO (FMVSS/CUR), motor vehicles and trailers are to be equippped with pneumatic tires having a minimum tread depth of ≥ 1.6 mm across their entire width at all points on the circumference.

Passenger vehicles and motor vehicles with an approved gross vehicle weight G of ≤ 2.8 metric tons designed to operate at a maximum speed v_{max} of > 40 km/h, and their trailers, can be equipped with either bias-ply or radial tires, the stipulation being that the same configuration be mounted at all wheels; in combinations, this stipulation applies to the individual vehicles only. It does not apply to trailers towed by a motor vehicle at speeds $v \leq 25$ km/h.

Table 1.
Tire categories and applicable standards

No	Tire application	German standards[1]	
		DIN	WdK[2]
1	Motor cycles	7801, 7802, 7810	119
2	Passenger cars	7803	128, 203
3	Light-duty commercial trucks	7804	132, 133
4	Commercial vehicles	7805, 7793	134, 135, 142, 143, 144, 153
5	Earth-moving machines	7798, 7799	145, 146
6	Industrial trucks	7811, 7845	171
7+8	Agricultural vehicles and machinery	7807, 7808, 7813	156, 161

[1] The corresponding European standards can be found in the "Data Book of Tires and Rims" of the ETRTO (European Tire and Rim Technical Organization, Brussels).
[2] Guideline of the Trade Association of the German Rubber Industry, Frankfurt.

Tire applications

Tire selection based upon the tire manufacturer's recommendations is essential for obtaining satisfactory performance. Optimal operating characteristics also depend upon tires of a single design (e.g., radial tires) being mounted on all wheels. To ensure maximum durability and safety, the tires should be maintained, stored and mounted in accordance with the instructions and guidelines issued by the tire manufacturer or tire specialist.

In actual use, viz., with the tires mounted on the vehicle, attention should be directed toward several factors. It is important to ensure that the tires are properly balanced – that is, that rotation is as concentric as possible. Tires of a single type should be used on all wheels, and the tire specification must correspond to the intended vehice application. Finally, the speed rating must be appropriate, and adequate tread depth is also essential.

Tread depth

Low tread depth is accompanied by a commensurate reduction in the size of the protective layer covering belts and carcass.

Tire rotation between axles is recommended when the tires at various axles are exposed to different rates of tread wear.

Adequate tread depth is especially important for operating safety on passenger cars and high-speed commercial vehicles, as worn tread leads to reduced traction on wet roads.

Reduced tread depth also leads to an overproportional increase in braking distances (Table 2).

Table 2. Tread depth and braking distance from 100 km/h

Tread	in	8	4	3	2	1
depth	mm					
Braking	in m	70	82	87	97	118
distance	in %	100	117	124	139	169
Increased braking distance in % per mm of wear		4		7	15	30

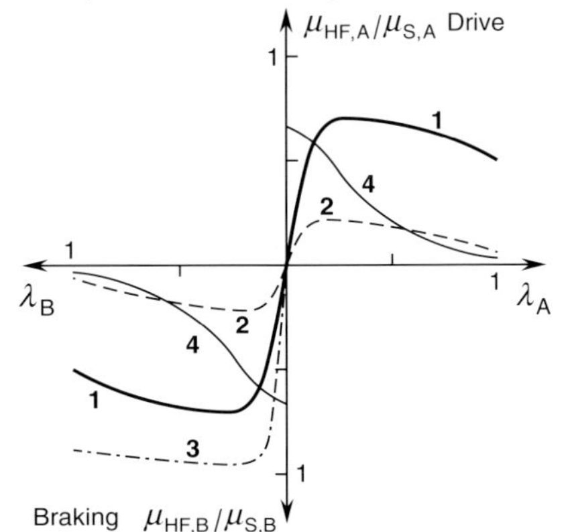

Fig. 3: Adhesion/slip curves
1 Wet asphalt $\mu_{HF, A/B}$,
2 Packed snow $\mu_{HF, A/B}$,
3 Concrete $\mu_{HF, B}$,
4 Asphalt $\mu_{S, A/B}$,
$\mu_{HF, A/B}$, Adhesion coefficient, acceleration/braking,
$\mu_{S, A/B}$ Lateral-force coefficient, acceleration/braking,
$\lambda_{A/B}$ Slip, acceleration/braking.

Slip

The amount of force which can be transferred during initial drive-off and during acceleration is (as for braking) a function of the amount of slip generated between the tire and the road surface. The adhesion/slip curves for acceleration and braking display the same basic pattern (Fig. 3).

The vast majority of acceleration and braking manoeuvres entail only limited amounts of slip, allowing response to remain within the stable range in the diagram; under these conditions, any rise in slip will be accompanied by a corresponding increase in available adhesion.

Further increases in slip take the curves through the maxima and into the instable range, where any further increase in slip will generally lead to reduced adhesion. A wheel will then lock within a few tenths of a second under braking, while the build-up of excess torque during acceleration leads to a rapid increase in rotational speed at one or both drive wheels.

When additional factors (such as high wheel loads or extreme wheel angles) are superimposed on the slip curve, the result is a further negative effect upon force-transfer and operating characteristics.

Dynamics of linear motion

Motive and braking forces

Acceleration and deceleration
The vehicle is regarded as accelerating or decelerating at a constant rate when the acceleration a (or deceleration $-a$) remains constant. The following equation applies where the initial or terminal velocity is zero:

$$a = \frac{v}{t} = \frac{2s}{t^2}$$

Because braking distances play such a vital role in road safety, the deceleration distance s is more critical than the distance travelled during acceleration.

Maxima for acceleration and braking
These maximum rates of acceleration and deceleration are obtained when the motive or braking forces at the vehicle's wheels reach such a magnitude that the tires are just within their traction limits (maximum coefficient of adhesion). The real-world figures are always somewhat lower, as the vehicle's tires do not simultaneously exploit their maximum adhesion during each acceleration (deceleration) process. Electronic traction and antilock braking systems (ASR, ABS) maintain the traction level in the vicinity of the maximum coefficient of friction.

Stopping distance

Definitions
This is the distance covered in the interval between the time when a hazard or obstacle is recognized and the point where the vehicle comes to a halt. It is the sum of the distance traveled during the reaction time t_r, the brake system's initial response delay t_a (at constant vehicle speed v) and the distance covered during the effective braking time t_w. Maximum retardation a is obtained during the pressure build-up period t_s. Alternatively, half of the pressure build-up period can

be viewed as representing full deceleration. The periods in which no active retardation occurs are combined to form the cumulative response delay, or time loss t_{vz} (Fig. 4):

$$t_{vz} = t_r + t_a + \frac{t_s}{2}$$

The upper limits on retardation are determined by the static coefficient of friction between tires and road surface, while the lower extremes are defined by the legally stipulated minima.
The difference between the stopping time t_h, or stopping distance s_h on the one hand, and the braking time t_b or braking distance s_b on the other, is defined by t_{vz} and $v \cdot t_{vz}$:

$$t_h = t_{vz} + \frac{v}{a}$$

$$s_h = v \cdot t_{vz} + \frac{v^2}{2a}$$

Reaction time
The reaction time is the period that elapses between recognition of the hazard or obstacle, the driver's decision to apply the

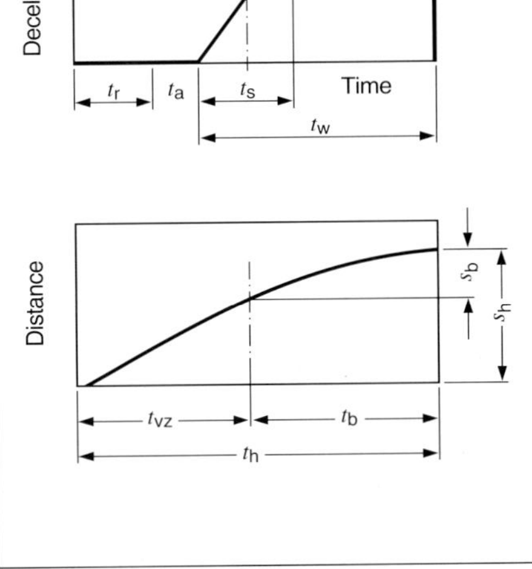

Fig. 4: Deceleration and stopping distances
s_b Braking distance. s_h Stopping distance, t_a Response time, t_b Braking time, t_h Stopping time, t_r Reaction time, t_s Pressure build-up time, t_{vz} Lost time, t_w Active braking time.

foot to the brake, and the time it takes for the foot to contact the brake pedal. The reaction time is not a fixed constant; depending upon the individual driver and various environmental variables, it can range from 0.3 to 1.7 seconds (Tables 3 and 4). Special tests (e.g., of the type conducted by medical and psychological institutes) are necessary to determine individual reaction patterns.

Brake response and pressure build-up times
The brake response and pressure build-up times t_a and t_s are defined by the brake system's control and transmission devices as well as by the instantaneous condition of the brakes themselves (e.g., wet brake rotors or discs). The legal regulations for tightly adjusted brakes are contained in ECE R13. EEC Regulation 71/320 is the reference used in defining brake-system effectiveness. It uses figures of 0.36 s (for Vehicle Category M_1) and 0.54 s (Vehicle categories M_2, M_3 and N_1 through N_3) in the braking-distance equation $t_a + t_s/2$ (refer to section on regulations). The response and pressure build-up times are longer if the brake system is in poor condition. A response delay of 1 s results in the stopping distances indicated in Table 5.

Passing (overtaking)
The passing distance s_u is the total distance covered when pulling out of the lane, overtaking the other vehicle, and then returning to the original lane. The

Table 3:
Personal factors influencing reaction time

	Psycho-physical reaction $\rightarrow$			$\leftarrow \mid \rightarrow$	Muscular reaction	$\rightarrow$
Perceived object (i.e. traffic sign)	Perceive	Comprehend	Decide	Incipient response	Move	Actuated control element (brake pedal)
	Acuity of vision	Perception and understanding	Mental processing	Loco-motor system	Personal response time	

Table 4:
Effect of personal and extraneous factors on reaction time

Factors	**Reduction** down to 0.3 s $\leftarrow$	$\rightarrow$ **Extension** up to 1.7 s
Personal factors	Trained, reflexive response	Inexperience, uncoordinated response
	Good frame of mind; optimal performance potential	Poor frame of mind, e.g. fatigue
	Highly skilled driver	Lower level of driving skill
	Youth	Advanced age
	Anticipation	Inattentiveness, distraction
	Physical and mental health	Health or psychic disorders
		Panic, alcohol
Extraneous factors Traffic situation	Uncomplicated, easily comprehended, familiar, predictable	Complicated, difficult to comprehend, unpredictable, rarely encountered
Type of external stimulus	Explicit, conspicuous	Equivocal, inconspicuous
Location of stimulus	Within the field of vision	At edge of field of vision
Type of control device	Logical control layout	Poor control layout

complete overtaking processes can occur under a wide range of conditions. Due to the complexity involved in defining all possible variations, the following is based on passing at a constant velocity and at a constant rate of acceleration.

Passing distance

It is possible to simplify the graphic representation (Fig. 5) by treating the passing distance s_u as the sum of the two components s_H and s_L while disregarding the addition travel required for the lane changes:

$$s_u = s_H + s_L$$

The distance s_H which the faster (overtaking) vehicle must cover relative to the (hypothetically stationary) slower (overtaken) vehicle is the sum of the vehicle lengths l_1 and l_2 together with the safety margins s_1 and s_2:

$$s_H = s_1 + s_2 + l_1 + l_2$$

During the passing time tu the slower vehicle covers the distance s_L, and the overtaking vehicle must also travel this distance to maintain the safety margin:

$$s_L = t_u \cdot v_L$$

Safety margin

The minimum safety margin corresponds to the distance covered during the lost time (response delay) t_{vz}. The standard figure for a lost time of $t_{vz} \approx 1$ s (velocity in km/h) is $0.3 \cdot v$ (in meters). A minimum of $0.5 \cdot v$ (meters) is recommended on secondary roads outside cities and townships.

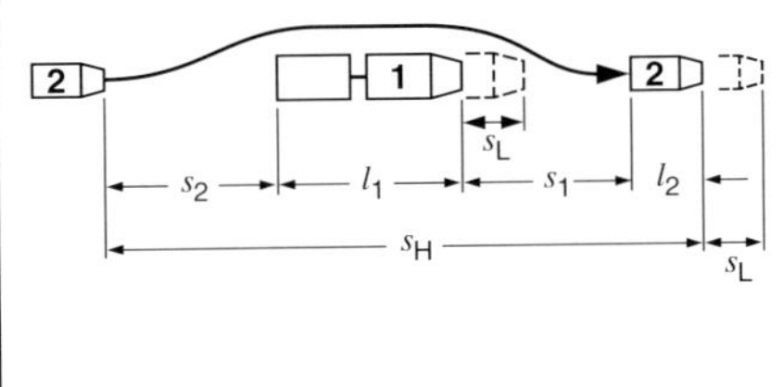

Fig. 5: Passing distance
l_1 and l_2 lengths of the vehicles concerned,
s_1 and s_2 safety margins,
s_H Relative distance travelled by overtaking vehicle,
s_L Distance travelled by vehicle being passed,
1 Vehicle being passed, 2 Passing vehicle.

Passing at a constant speed

On multi-lane highways, the overtaking vehicle will frequently be traveling at a speed adequate for passing even before the actual process starts. The passing time (from the initial lane change until the return to the original lane is completed) is then

$$t_u = \frac{s_H}{v_H - v_L}$$

with a corresponding passing distance

$$s_u = t_u \cdot v_H$$

Passing with constant acceleration

On narrower roads, the overtaking vehicle will usually have to decelerate to the speed of the preceding vehicle before again accelerating to pass. Available acceleration rates are determined by such factors as engine output, vehicle

Table 5:
Stopping distance s_h for a delay of 1 s

Deceleration a in m/s	Vehicle speed v in km/h prior to braking												
	10	30	50	60	70	80	90	100	120	140	160	180	200
	Distance travelled t_{vz} in m during delay of 1 sec (no braking)												
	2.8	8.3	14	17	19	22	25	28	33	39	44	50	56
	Stopping distance s_h in m												
4.4	3.7	16	36	48	62	78	96	115	160	210	270	335	405
5	3.5	15	33	44	57	71	87	105	145	190	240	300	365
5.8	3.4	14	30	40	52	65	79	94	130	170	215	265	320
7	3.3	13	28	36	46	57	70	83	110	145	185	230	275
8	3.3	13	26	34	43	53	64	76	105	135	170	205	250
9	3.2	12	25	32	40	50	60	71	95	125	155	190	225

weight, speed and running resistance. The rate generally lies within the range of $a = 0.4 \ldots 0.8$ m/s², with up to $a = 1.4$ m/s² available in lower gears for further reductions in passing time. The distance required to complete the passing manoeuvre should never exceed half the visible stretch of road.

Operating on the assumption that a constant rate of acceleration can be maintained for the duration of the passing manoeuvre, the passing time will be

$$t_u = \sqrt{\frac{2 \cdot s_H}{a}}$$

The distance which the slower vehicle traverses in this period is defined as $s_L = t_u \cdot v_L$. This gives a passing distance of:

$$s_u = s_H + t_u \cdot v_L$$

Range of visions

For safe passing on narrow roads, the range of visions must be at least the sum of the passing distance plus the distance that would be traveled by any oncoming vehicle while the passing manoeuvre is in progress. This distance is approximately 400 meters if the vehicles approaching each other are traveling at 90 km/h and the vehicle being overtaken is underway at 60 km/h.

Surface adhesion

Static coefficient of friction

The static coefficient of friction (tire-road interface friction coefficient) is defined by such factors as vehicle speed, tire condition and the state of the road surface. The figures in Table 6 apply to concrete and tarmacadam surfaces in good condition. The coefficient of sliding friction (with wheel locked) is generally lower than the coefficient of static friction.

Aquaplaning

Aquaplaning exercises a particularly dramatic influence on the contact between tire and road surface. This term refers to a state in which a layer of water separates the tire from the wet road surface. This phenomenon occurs when a wedge of water forms beneath the tire's contact patch, lifting it from the road. The tire starts to "swim".

The tendency for aquaplaning to occur is influenced by:
- the depth of the water on the road surface,
- the vehicle speed,
- the tread pattern, and
- tire wear, as well as
- the force with which the tire is being pressed against the road surface.

Wide tires are particularly susceptible to aquaplaning. It is impossible to steer or brake an aquaplaning vehicle, as neither steering inputs nor braking forces can be transmitted to the road surface (Fig. 6).

Table 6:

Coefficients of static friction μ_{HF} for pneumatic tires on various road surfaces

Vehicle speed	Tire conditions	Road conditions Dry	Road conditions Wet (water approx. 0.2 mm deep)	Heavy rainfall (water approx. 1 mm deep)	Puddles (water approx. 2 mm deep)	Ice (black ice)
km/h		μ_{HF}	μ_{HF}	μ_{HF}	μ_{HF}	μ_{HF}
50	New	0.85	0.65	0.55	0.5	0.1 and less
	Worn	1	0.5	0.4	0.25	
90	New	0.8	0.6	0.3	0.05	
	Worn	0.95	0.2	0.1	0.05	
130	New	0.75	0.55	0.2	0	
	Worn	0.9	0.2	0.1	0	

Total running resistance

The total running resistance F_W is defined as the sum of the rolling resistance F_{Ro}, the aerodynamic drag F_L and climbing resistance F_{St}:

$$F_W = F_{Ro} + F_L + F_{St}$$

A corresponding force must be applied through the drive wheels to overcome this cumulative resistance to motion (Fig. 7).

Rolling resistance

The rolling resistance F_{Ro} is the result of the deformation processes which occur at the contact patch between tire and road surface, and is defined as the product of the gravitational force G and the coefficient of rolling resistance f:

$$F_{Ro} = f \cdot G$$

The coefficient of rolling resistance is directly proportional to tire deformation, and inversely proportional to the tire radius. The coefficient thus increases in response to greater loads, higher speeds, and lower tire pressures.

Examples: Passenger-car tires on

Concrete, asphalt	$f = 0.013$
Small sett pavement	$f = 0.015$
Rolled gravel	$f = 0.02$
Tarmacadam	$f = 0.025$
Unpaved road	$f = 0.05$

During cornering, the rolling resistance is augmented by the cornering resistance. The coefficient of cornering resistance is defined by such factors as vehicle speed, cornering radius, suspension geometry, tires, tire pressures and the vehicle's response patterns under wheel-slip conditions.

Aerodynamic drag

The factors used to determine the aerodynamic drag F_L are atmospheric density ϱ, the drag coefficient c_W (a function of vehicle design), the maximum vehicle section A and the vehicle speed v (including headwind velocity v_0):

$$F_L = 0.5 \cdot \varrho \cdot c_W \cdot A \, (v + v_0)^2$$

Drag coefficients for various automotive configurations:

Open convertible	$c_W = 0.5...0.7$
Station wagon	$c_W = 0.5...0.6$
Conventional form[1]	$c_W = 0.4...0.55$
Wedge shape	$c_W = 0.3...0.4$
Enclosed	$c_W = 0.2...0.25$
Teardrop	$c_W = 0.15...0.2$

Drag coefficients for trucks:
Conventional tractors

– no aerodynamic aids	$c_W \geq 0.64$
– some aids	$c_W = 0.54...0.63$
– fully equipped	$c_W \leq 0.53$

---

[1] Conventional 3-box design

Fig. 6: Aquaplaning
1 Tire, 2 Wedge of water, 3 Road surface.

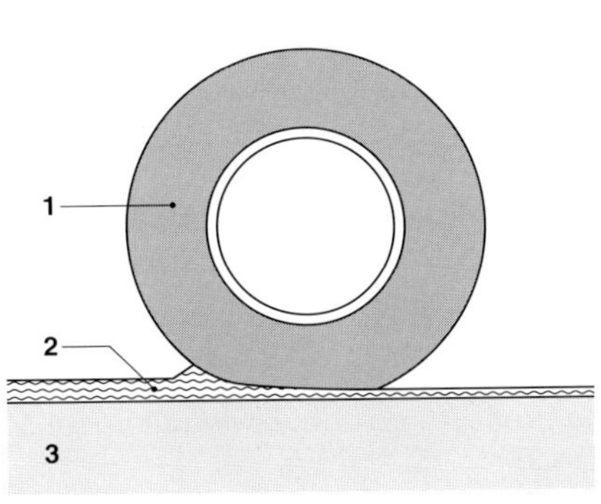

Fig. 7: Total running resistance
F_L Aerodynamic drag, F_{Ro} Rolling resistance, F_{St} Climbing resistance, S Center of gravity, G Weight, α Gradient angle.

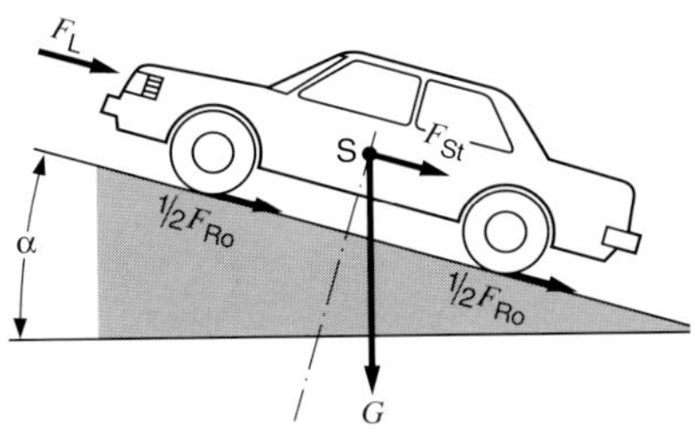

Climbing resistance

The climbing resistance F_{St} (with positive operational sign) and the downgrade force (with negative sign) are functions of gravitational force exerted on the vehicle's mass times the gradient angle:

$$F_{St} = G \cdot \sin \alpha$$

Motive force

The motive force F available at the drive wheels increases in response to higher levels of engine torque M and higher overall conversion ratios i between the engine and the drive wheels. It decreases in response to force-transmission losses (efficiency η with longitudinal engine approximately 0.88...0.92, with transverse engine approximately 0.91...0.95); r dynamic tire radius:

$$F = M \cdot i \cdot \frac{\eta}{r}$$

This motive force is partially consumed in overcoming the running resistance. Numerically higher conversion ratios are employed to overcome the substantially higher levels of running resistance encountered when ascending gradients (variable-speed transmission/gearbox).

Acceleration

The surplus force $F - F_W$ remaining when running resistance is subtracted from motive force accelerates the vehicle. The rotational inertia coefficient k_m defines the hypothetical increase in vehicle mass m due to rotating masses (wheels, flywheel, crankshaft, etc):

$$a = \frac{F - F_W}{k_m \cdot m}$$

If total running resistance is in excess of motive force the vehicle responds by decelerating.

Dynamics of lateral motion

Response to crosswinds

Strong crosswinds can deflect the vehicle from its intended path of travel, and can be an especially critical factor at high vehicle speeds and with unfavorable vehicle dimensions. Sudden sidewinds of the kind encountered when embankments give way to open, unsheltered roads can evoke yaw angle changes as well as causing the vehicle to drift off course. As these processes can take effect in less than the driver's response time, they can also provoke false reactions on the part of the vehicle's operator.

When wind impacts upon a vehicle from an oblique angle, aerodynamic drag is supplemented by a lateral aerodynamic force or component. This force – known as the crosswind force – can be viewed as being distributed across the entire body of the vehicle, but it is concentrated at the "pressure point." The exact location of this point is determined by the body configuration and the angle of attack of the wind.

The pressure point is usually found in the forward half of the vehicle. On conventional ("three-box") bodies, it tends to be found – with a high degree of consistency – closer to the center of the vehicle than on hatchback (sloping tail) vehicles, in which the pressure point displays a tendency to shift in response to variations in wind angle. In contrast, the position of the center of gravity S varies according to vehicle load factor. As an aid to representing the influence of crosswinds (independent of the relative positions of bodywork and suspension) a reference point O is defined at the center of the vehicle in its front half.

When the crosswind force F_S is defined for a reference point that does not coincide with the pressure point, it is supplemented by the crosswind moment at the pressure point – yawing moment M_Z (Fig. 8).

The crosswind force is taken up by the lateral forces generated at the tires. The degree of lateral force generated by a vehicle tire depends upon numerous factors, includ-ing the wheel-slip angle, wheel load, tire design and dimensions, its inflation pressure and the frictional properties of the road surface.

In the interests of directional stability in crosswinds, the pressure point should be located in the vicinity of the vehicle's center of gravity. If the vehicle's basic handling attitude is characterized by oversteer, the vehicle will demonstrate a mild tendency to veer if the pressure point is forward of the center of gravity. On understeering vehicles the optimal position for the pressure point is just to the rear of the center of gravity.

Oversteer and understeer

Cornering forces can only be generated between a rubber tire and the road surface on which it is rolling when the tire is at a planar angle; in other words, a certain wheel-slip angle is required.

A vehicle is defined as understeering when it responds to increasing rates of lateral acceleration by generating higher slip angles at the front than at the rear. The inverse response pattern is defined as oversteer.

Some vehicles display consistent oversteer or understeer. Others, however, may respond with initial understeer at low lateral accelerations before shifting to oversteer at higher rates (or vice versa).

Centrifugal force in curves

The centrifugal force F_{cf}, with the center of gravity S as its focal point, is determined by such factors as vehicle mass m, velocity v and the cornering radius r_K (Fig. 9):

$$F_{cf} = m \cdot \frac{v^2}{r_K}$$

Roll axis

During cornering, the centrifugal force emanating from the center of gravity tends to deflect the vehicle from its original path. The magnitude of this tendency depends on the spring rates (with varying compression on different sides of the vehicle) and upon the lever arm of centrifugal force (distance between the center of gravity and the roll axis). The roll axis is the body's instantaneous axis of rotation relative to the road surface. As a rigid body in space, the body consistently executes a screwing or rotating motion; this motion is supplemented by a lateral displacement along the instantaneous axis.

A higher roll axis – that is, the closer it is to a parallel axis through the center of gravity – results in higher transverse stability, and lower roll tendency.

Fig. 8: Vehicle exposed to crosswind
D Pressure point, O Reference point, S Center of gravity.
F_S Crosswind force, M_z Yawing moment.
$F_S + M_z$ impacting at O corresponds to F_S acting at D.
v_r Air-flow velocity,
v_s Crosswind velocity,
v_w Headwind velocity, τ Air-flow angle.

Fig. 9: Centrifugal force when cornering
b Trackwidth, g Acceleration due to gravity,
h_s Height of center of gravity, m Vehicle mass,
r_K Curve radius, G Weight, F_{cf} Centrifugal force,
β Road camber, S Center of gravity.

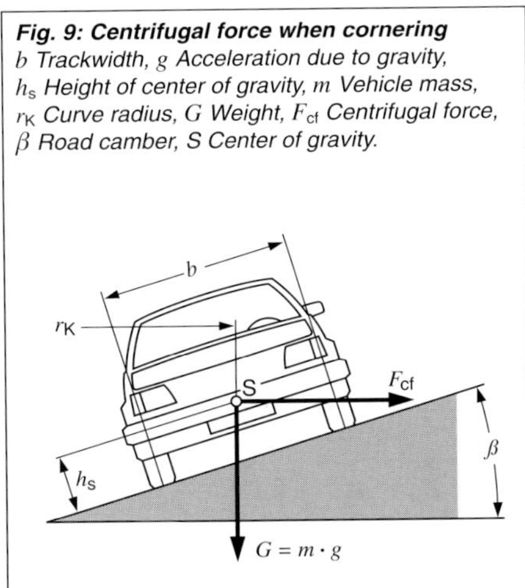

Brake systems
for passenger cars

Brake systems

Passenger-car brake systems can be classified according to the following criteria:
- design concepts, and
- operating principles.

Design concepts

Based on official regulations, the automotive braking equipment's functions can be divided into three brake systems:
- service-brake system (BBA),
- secondary-brake system (HBA), and
- parking-brake system (FBA).

Service-brake system
The service brakes (footbrake) can be employed to reduce the vehicle's speed, to maintain it at a constant level (for instance on a gradient) and to bring it to a halt. This is the system employed in the course of normal operation. It provides precisely-controlled, variable braking response at all four wheels.

Secondary-brake system
In the event that the service brakes fail, the secondary-brake system must be capable of assuming its functions, although it may generate only reduced braking force.
The secondary (or auxiliary) brake system need not necessarily consist of a separate third system (supplementing the service and parking brakes) with its own control mechanism; it may also comprise the intact circuit in a dual-circuit service-brake layout or of a parking brake capable of graduated response.

Parking-brake system
The parking-brake (handbrake) system assumes the third braking function. It must be capable of maintaining the vehicle in a stationary state, even on gradients and in the absence of the driver.
Safety considerations dictate that the parking-brake system feature a continuous mechanical connection between the control mechanism and the wheel brake, e.g., linkage rods or a Bowden cable.
The parking brake is actuated from the driver's seat, in most cases using a hand lever, in others via a pedal. This brake system is designed to provide graduated response. It operates on the wheels at one axle only.

Operating principles

Depending upon whether a brake system is operated completely, partially or not at all by muscular energy, a distinction is drawn between:
- muscular-energy brake systems,
- power-assisted brake systems, and
- power-brake systems.

Muscular-energy brake systems
This type of system is installed in passenger cars and two-wheeled vehicles. The muscular force applied at the pedal or hand lever is transmitted to the brakes via a mechanical (linkage rods or Bowden cable) or hydraulic (master cylinder, wheel cylinders) relay system.

Power-assisted brake systems
The power-assisted (energy-assisted) brake system is found in passenger cars and light commercial vehicles. This type of unit employs a brake booster (servo unit) to supplement muscular force with

energy generated by vacuum or hydraulic pressure. A hydraulic circuit then transmits this enhanced muscular force to the wheel cylinders (Fig. 1).

Power-brake systems

The chief field of application for this non-muscular-energy brake system is in heavy commerical vehicles, but this type of system is also occasionally found in large passenger vehicles with integral antilock braking systems (ABS). With this system, the force used to operate the service brakes is entirely non-muscular.

These systems employ hydraulic energy (based on hydrostatic pressure) and hydraulic transmission devices The hydraulic fluid is stored in energy reservoirs (hydraulic accumulators) containing a compressed gas (usually nitrogen).

A flexible diaphragm or bladder, or, alternatively, a piston with a rubber seal (piston accumulator) is employed to separate the fluid from the gas.

The hydrostatic pressure, which remains in a constant state of balance with the gas pressure, is generated by a hydraulic pump. A pressure regulator switches the hydraulic pump back to idle whenever maximum pressure is reached.

One of the advantages of hydraulic fluid is that it may be viewed as incompressible in normal use – its volume remains constant regardless of pressure. This makes it possible to use small quantities of hydraulic fluid to transmit high levels of pressure, and the master cylinder retains its compact dimensions.

It is even possible to install ABS in the system without substantially increasing its complexity, as the pressure-release process at the wheel cylinders can be accomplished by discharging hydraulic fluid back into the supply reservoir (hydraulic accumulator).

A liability of this design resides in the fact that any system leak will lead to a loss of hydraulic fluid, ultimately culminating in complete depletion of the system's energy-transfer medium.

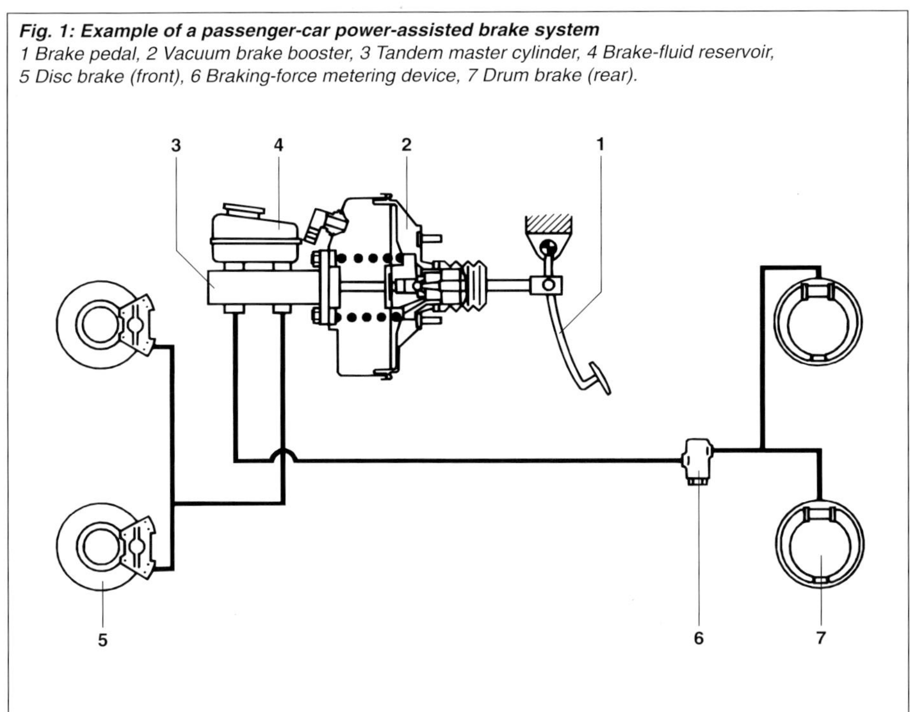

Fig. 1: Example of a passenger-car power-assisted brake system
1 Brake pedal, 2 Vacuum brake booster, 3 Tandem master cylinder, 4 Brake-fluid reservoir, 5 Disc brake (front), 6 Braking-force metering device, 7 Drum brake (rear).

Friction forces

The friction force F_R is proportional to the normal force F_N:

$$F_R = \mu_{HF} \cdot F_N$$

The factor μ_{HF} is the braking-force coefficient (or frictional, or positive-adhesion coefficient). It describes the characteristics of the various tire/road-surface materials combinations as well as the influences to which these combinations are exposed. The braking-force coefficient serves as an index for the amount of braking force that can be applied. The highest frictional coefficients for automotive tires are obtained on dry, clean road surfaces; the lowest occur on ice. Intermediate states are produced by other substances, such as water and dirt, which also reduce the coefficient of friction.

Example:

Road surface condition	Braking force coefficient μ_{HF}
dry	0.8 ... 1
wet	0.2 ... 0.65
icy	0.05 ... 0.1

Speed is always an important factor for the braking-force coefficient, but its role is particularly dramatic on wet road surfaces.

Braking initiated at high speeds on less than ideal highway surfaces can lead to wheel lock whenever the braking-force coefficient is too low to ensure that the vehicle's wheels will adhere to the road surface. Once a wheel locks, it looses the ability to transfer lateral forces, and the vehicle ceases to respond to steering inputs. Figure 6 illustrates the frequency distributions for braking-force coefficients

Fig. 4: Wheel rolling movement
a) Freely rolling wheel, b) Braked wheel,
v_F *Vehicle speed at the wheel center point M,*
v_U *Peripheral wheel speed.*
Braking the wheel reduces the angle of rotation φ per unit of time (slip).

a

b

Fig. 5: Tire conditions when cornering
α *Slip angle, F_S Lateral force,*
v_S *Wheel's center-of-gravity velocity,*
m *Wheel center plane.*

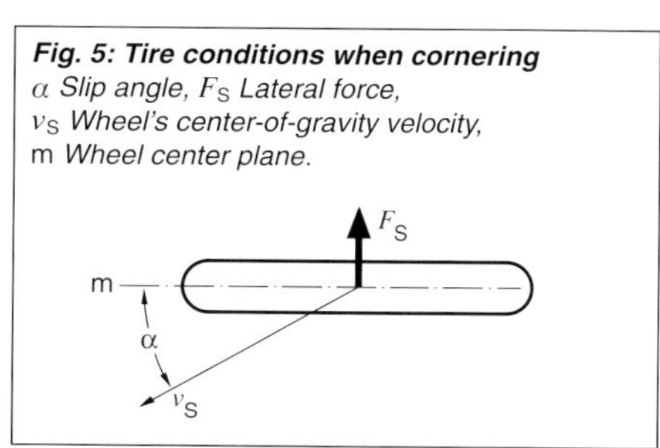

Fig. 6: Braking-force coefficient: Frequency distribution on a braked wheel at various speeds on wet road surfaces.
(Source: Forschungsinstitut für Kraftfahrwesen und Fahrzeugmotoren in Stuttgart).

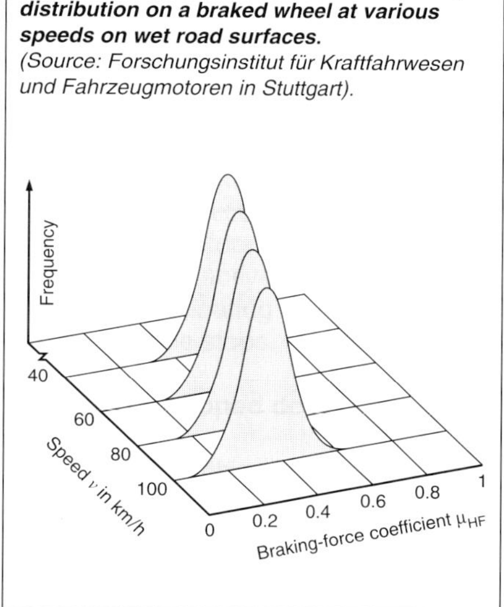

for a locked wheel at various speeds on wet roads.

Frictional processes are divided into two subcategories: static and sliding friction. For rigid bodies, the static friction is greater than the sliding friction. The corollary is that conditions exist in which a moving rubber tire will generate a higher coefficient of friction than those that it produces when locked. At the same time, sliding processes also occur as the tire rolls during normal operation. This condition is called "slip."

Slip

As the wheel turns in response to accelerative or braking forces, its motion is accompanied by complex physical processes at the tire contact patch. The rubber elements are subject to tension and distortion, and partial slip also occurs below the wheel-lock threshold. The slip factor λ indicates the proportion of slip in the rolling process:

$$\lambda = (v_F - v_U)/v_F$$

where v_F is vehicle speed and v_U the tire's peripheral velocity (Fig. 4). As the equation indicates, braking slip occurs as soon as the wheel decelerates to a rotational speed below that which would normally correspond to a given vehicular velocity. This is the only state in which braking forces can be generated (an analogous situation is encountered during acceleration). The coefficients of friction – frequently portrayed as a function of braking slip – can vary according to road = surface conditions. When the brakes are applied during linear operation, no lateral forces are generated, so the entire traction potential at the contact patch between tire and road is available for braking. From an initial braking slip of zero, the braking-force coefficient μ_{HF} rapidly rises to a maximum figure of between 10 % and 40 % (the exact figure varies according to tire and road conditions) before again decreasing.

The rising sections of the curves are called the stable range (partial braking range), and the falling section is the

instable range (Fig. 7). During linear braking, ABS prevents the vehicle from entering this instable range.

To ensure that the vehicle remains on track during cornering, lateral forces must be generated at the tires to counteract the centrifugal force as it seeks to move the vehicle outward from its center of gravity.

A precondition for generating lateral forces is lateral deformation in the tires. In this state, the velocity at the wheel's center of gravity v_s deviates from the center plane "m" by the slip angle (Fig. 5).

Fig. 7 illustrates the braking-foce and lateral-force coefficients as functions of the brake slip at a slip angle of 4°. The coefficient of lateral force rises to its maximum at a brake slip of zero. It then responds to increasing levels of brake slip by dropping, slowly at first, then more rapidly; the bottom of the curve is reached when the wheel locks. This minimum value is due to the slip angle of the locked wheel and the fact that this wheel is no longer capable of generating lateral resistance.

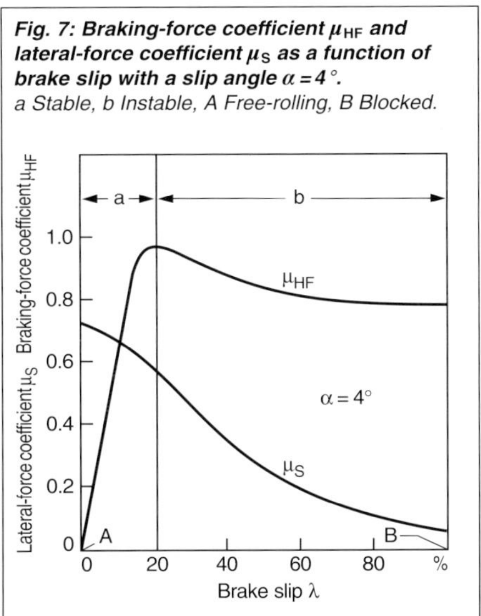

Fig. 7: Braking-force coefficient μ_{HF} and lateral-force coefficient μ_S as a function of brake slip with a slip angle $\alpha = 4$°.
a Stable, b Instable, A Free-rolling, B Blocked.

Brake-system components

Braking joins steering and shifting gears as one of the most frequently executed driver operations. Accordingly, the brake system's components must be designed to extract maximum use from the applied foot pressure and/or maintain the corresponding pressure requirements at the lowest possible level.

Brake booster

The brake booster amplifies the foot pressure applied when actuating the brakes, and in doing so reduces the manual effort required to operate them. In most automotive braking systems the brake booster is found in a combined assembly with the master cylinder. It is essential that the brake booster should not impair precise, sensitive control of braking force. Two types of brake booster are in general use. Both the vacuum and the hydraulic brake booster operate by exploiting an energy source that is already available in the vehicle.

Vacuum brake booster

The vast majority of passenger-car brake systems are equipped with a vacuum-type brake booster (Fig. 1).

The vacuum brake booster uses the negative pressure generated in the SI engine's intake tract – or the force produced by the vacuum pump on cars with diesel engine (0.5...0.9 bar) – to amplify the force produced at the pedal. When the brakes are applied, this supplementary force increases as a direct function of pedal force, and continues to increase until it reaches the "cycle" pressure. This point, which lies in the vicinity of the locking point for the front wheels, is between 60 and 100 bar, depending upon the individual vehicle. There are no further increases in boost pressure beyond this point.

Fig. 1: Vacuum brake booster
1 Push rod (output force to tandem master cylinder), 2 Vacuum chamber with vacuum connection, 3 Diaphragm, 4 Working piston, 5 Double valve, 6 Air filter, 7 Piston rod (pedal force), 8 Working chamber.

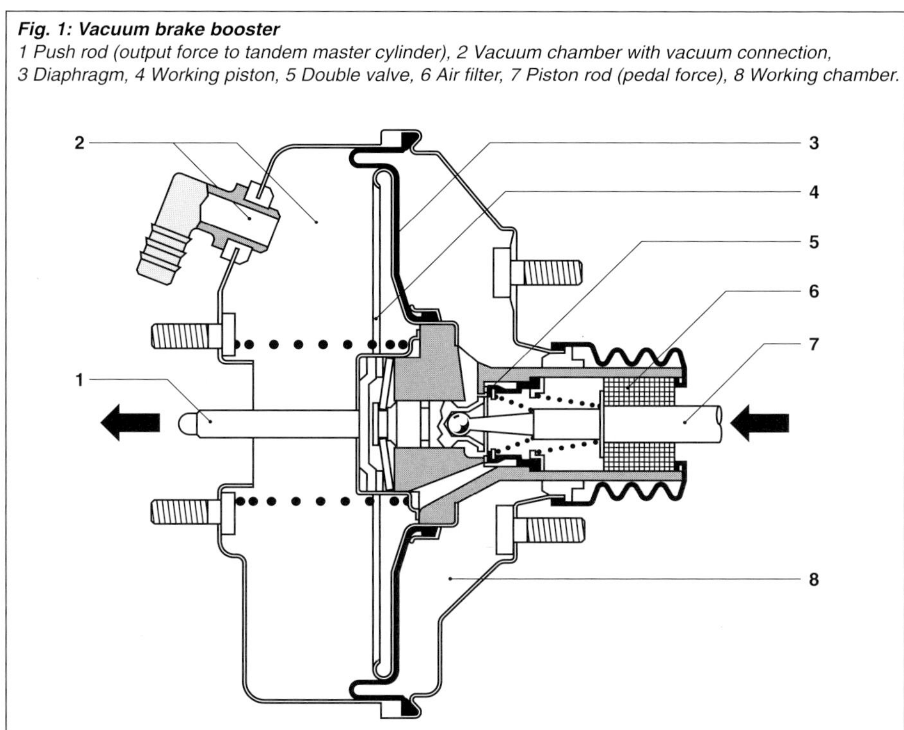

Hydraulic brake booster

This type of brake booster can be installed on vehicles which generate only minimal intake vacuum (e.g., diesel and turbo-engines) and which are already equipped with a hydraulic energy source (e.g., for the power steering). The hydraulic brake booster is much more compact and features a higher cycle pressure (in the range of 160 bar) than the vacuum brake booster.

One liability is the generally "spongy" pedal response.

Master cylinder

The braking process is initiated and controlled through the master cylinder. Official regulations stipulate that passenger cars shall be equipped with two separate brake circuits. This requirement is satisfied by using a master cylinder designed as a tandem unit (Fig. 2). The float piston (6) responds to leakage in the secondary or "floating" circuit by traveling to the end of the cylinder, allowing pressure to accumulate in the pressure chamber (21). If a leak occurs in the primary, or "rod" circuit, the push-rod piston (14) shifts against the float piston (6), pushing it to the left to permit pressure to accumulate. When the brakes are applied, extended travel and increased force levels required at the brake pedal will alert the driver to the fact that one of the circuits has failed. The master cylinder described here is also equipped with a central valve in the floating circuit; when pressure is released, the brake fluid flows through the orifice in the valve pin (18). Another passage serves as a permanent connection between the intermediate chamber (8) and the supply reservoir. An advantage of using a central valve is that it makes the snifter bore (12) redundant. On ABS-equipped vehicles, there is a risk of damage to the primary cup seal (23) when the piston passes over the snifter bore (12) at high pressure (leading to brake-circuit failure). This is why the master cylinders on most ABS-equipped vehicles feature two central valves.

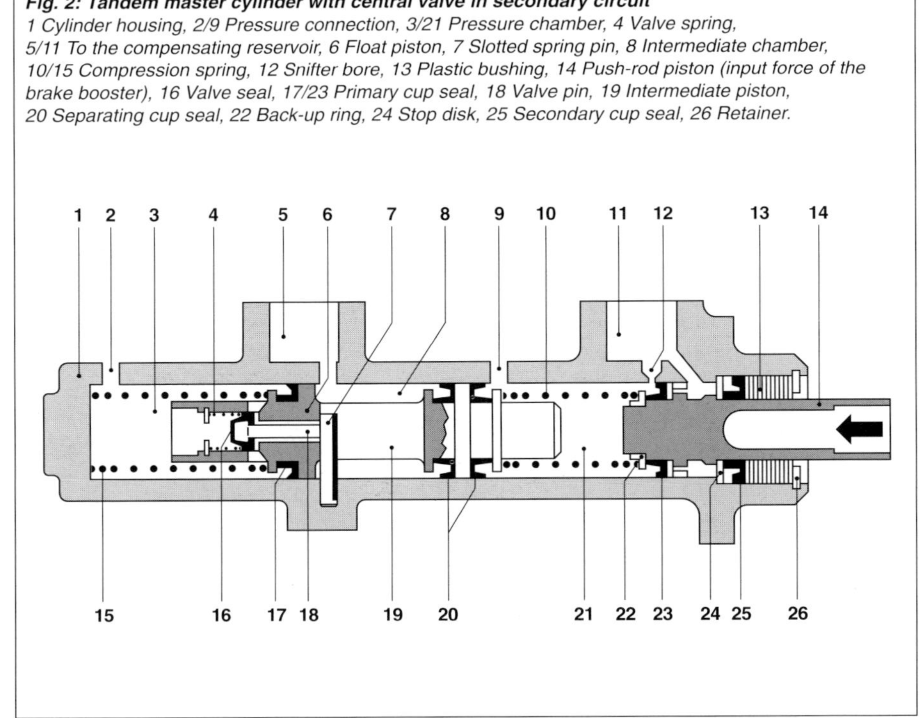

Fig. 2: Tandem master cylinder with central valve in secondary circuit
1 Cylinder housing, 2/9 Pressure connection, 3/21 Pressure chamber, 4 Valve spring,
5/11 To the compensating reservoir, 6 Float piston, 7 Slotted spring pin, 8 Intermediate chamber,
10/15 Compression spring, 12 Snifter bore, 13 Plastic bushing, 14 Push-rod piston (input force of the brake booster), 16 Valve seal, 17/23 Primary cup seal, 18 Valve pin, 19 Intermediate piston, 20 Separating cup seal, 22 Back-up ring, 24 Stop disk, 25 Secondary cup seal, 26 Retainer.

Operation

The force applied to the brake pedal acts directly on the push-rod piston (14), which reacts by moving to the left. As it does so, it passes over the snifter bore (12), and the fluid in the pressure chamber (21) can press the float piston (6) to the left as well.

The valve pin (18) is no longer up against the slotted spring pin (7) as soon as the floating piston has moved about 1 mm to the left. The valve seal (16) presses against the float piston (6) to seal and insulate the pressure chamber (3) from the intermediate chamber (8).

The pressure in both chambers (3 and 21) will now increase in response to any additional force at the pedal. At the same time, both pistons (6 and 14) will react to decreasing foot pressure by shifting to the right until the snifter bore (12) is exposed or until the valve pin (18) contacts the slotted spring pin (7) and lifts the valve seal (16) from the float piston (6). This allows the brake fluid to return to the supply reservoir, releasing the pressure in the brake system.

Brake-pressure regulating valves

Due to the shift in dynamic forces from rear to front that accompanies vehicle braking, the braking force applied to the front wheels must be greater than that at the rear. This is why front-brake components are larger than those at the rear wheels. However, this rear-to-front weight shift is not a linear process. Its magnitude increases as a function of deceleration. Thus the need for a supplementary means of reducing the brake pressure at the rear wheels relative to that at the front wheels. This function is assumed by the brake-pressure regulating valve. Depending upon the type of vehicle and the systems employed by the individual manufacturer, this pressure regulating valve will usually conform to one of three designs; these are the

– load-sensitive pressure-regulating valve,
– pressure-sensitive pressure-regulating valve, and
– deceleration-sensitive pressure-regulating valve.

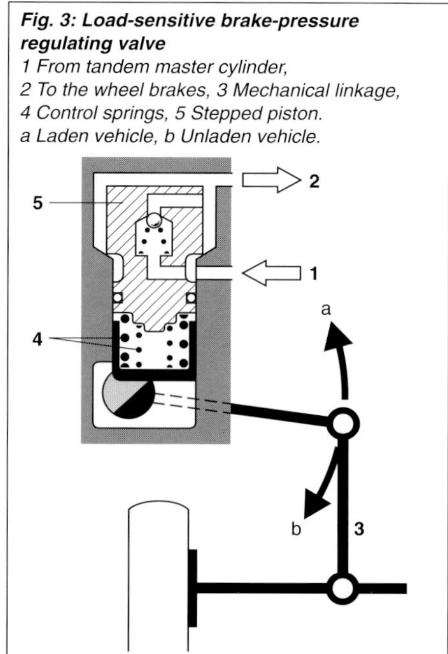

Fig. 3: Load-sensitive brake-pressure regulating valve
1 From tandem master cylinder,
2 To the wheel brakes, 3 Mechanical linkage,
4 Control springs, 5 Stepped piston.
a Laden vehicle, b Unladen vehicle.

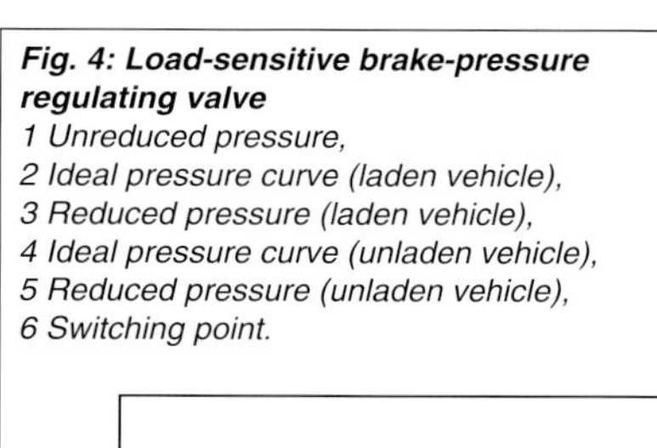

Fig. 4: Load-sensitive brake-pressure regulating valve
1 Unreduced pressure,
2 Ideal pressure curve (laden vehicle),
3 Reduced pressure (laden vehicle),
4 Ideal pressure curve (unladen vehicle),
5 Reduced pressure (unladen vehicle),
6 Switching point.

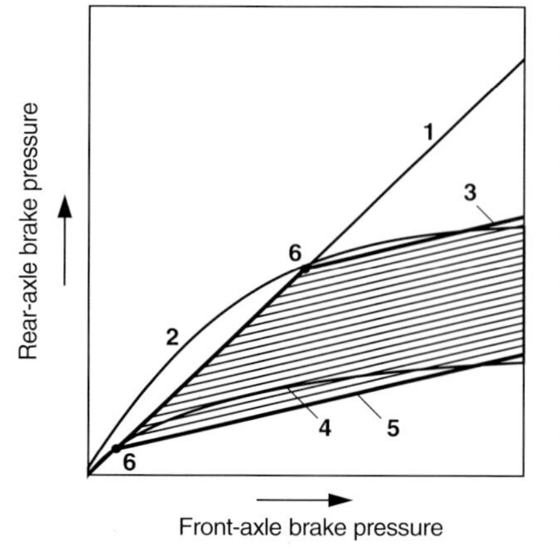

Load-sensitive brake-pressure regulating valves

The load-sensitive pressure-regulating valves are required on those vehicles (such as station wagons) in which high load factors lead to pronounced inter-axle force shifts during braking (Fig. 3 and 4). The pressure regulator is attached to the vehicle body, and connected to the rear suspension via mechanical linkage. The relative displacement of suspension and body is relayed to a piston located within the pressure-regulating valve's housing. This piston responds to variations in suspension compression rates by contracting a spring to modify the cycle point. This system adapts the pressure at the rear brakes to compensate for variations in load conditions.

Pressure-sensitive brake-pressure regulating valves

Pressure-sensitive pressure-regulating valves, also known as pressure limiters (Fig. 5), are employed on vehicles in which the potential for inter-axle load shifts is restricted by limited cargo capacity and a low center of gravity (for instance, sports cars).

Deceleration-sensitive brake-pressure regulating valves

Deceleration-sensitive pressure-regulating valves (Fig. 6) are used in a wide variety of applications. The cycle point in these devices, determined by the vehicle's rate of deceleration, is generally 0.3 g (with g as gravitational acceleration). This valve incorporates effective compensation properties. Since the brake pressure required to maintain any given rate of deceleration will depend upon the vehicle's load, this valve provides load-sensitive operation. A liability of this design lies in the operational anomalies that may be encountered under certain kinds of braking.

Design

In designing a brake-pressure regulating valve, it is important to ensure that the <u>actual</u> brake-force distribution will remain consistently below the <u>theoretical ideal</u>. Other influences to be considered in preventing overbraking at the rear wheels are fluctuations in frictional coefficients, engine braking torque and the tolerances in the pressure regulator itself.

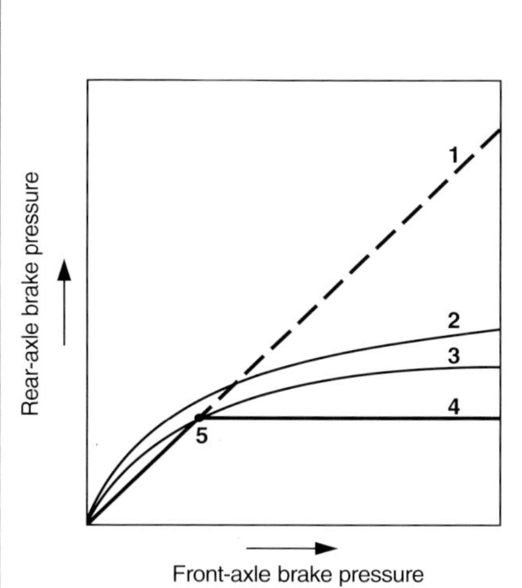

Fig. 5: Pressure-sensitive brake-pressure regulating valve
1 Unreduced pressure,
2 Ideal pressure curve (laden vehicle),
3 Ideal pressure curve (unladen vehicle),
4 Reduced pressure, 5 Switching point.

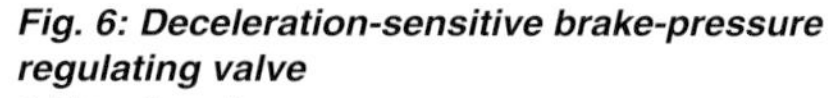

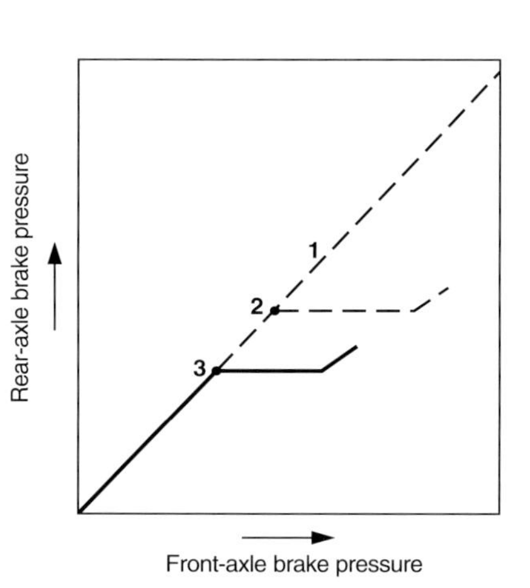

Fig. 6: Deceleration-sensitive brake-pressure regulating valve
1 Unreduced pressure,
2 Switch point (laden vehicle),
3 Switch point (unladen vehicle).

Wheel brakes

The operative distinction in wheel brakes is between discs and drums. Virtually all front brakes are disc units, and the trend is toward also installing discs at the rear. These are friction brakes in which the braking energy transmitted by the system operates as a tensioning force to press the brake pads (or linings) against the brake discs (or drums). The brake disc is also known as the brake rotor.

The wheel brakes must satisfy stringent requirements:
- short stopping distances,
- minimum response delay, and
- minimum build-up times before maximum effectiveness is achieved.

These requirements are satisfied in equal measure by disc and drum brakes. On vehicles equipped with 4-wheel disc brakes a supplementary drum unit is installed in the rear rotor hubs for the parking brake.
As the available deceleration rates should be maintained under continuous braking and repeated stops from high speeds, the wheel brakes must meet three basic requirements:

- adequate thermal absorption and dissipation,
- sufficient air flow over the brakes to dissipate the heat generated during braking, and
- the brake pads must maintain their frictional properties throughout an extended temperature range.

Disc brakes are superior to drum units in all three of these areas, and are therefore the configuration of choice in most applications.

Disc brakes
The braking forces on disc brakes are generated at the surface of a disc (or rotor) which rotates along with the vehicle's wheel, while the U-shaped caliper is supported by stationary vehicle components.

Fixed-caliper disc brakes
In each half section of the fixed caliper (Fig. 7) is a piston to which hydraulic pressure is applied during braking. Each of the pistons presses a brake pad against its respective side of the brake disc.
When the brakes are released, specially-shaped piston seals with programmed deformation properties retract the piston by a defined increment (approx. 0.2 mm). Thus no adjustments are required with disk brakes.
Due to their high levels of physical strength, fixed-caliper brakes are often installed on heavy and high-speed passenger cars. The liabilities of this design include thermal sensitivity under extended application (e.g., extended descents in alpine driving); complete brake failure from overheated brake fluid (see section on "Brake fluids") is more common with fixed-caliper than with floating-caliper disc brakes.

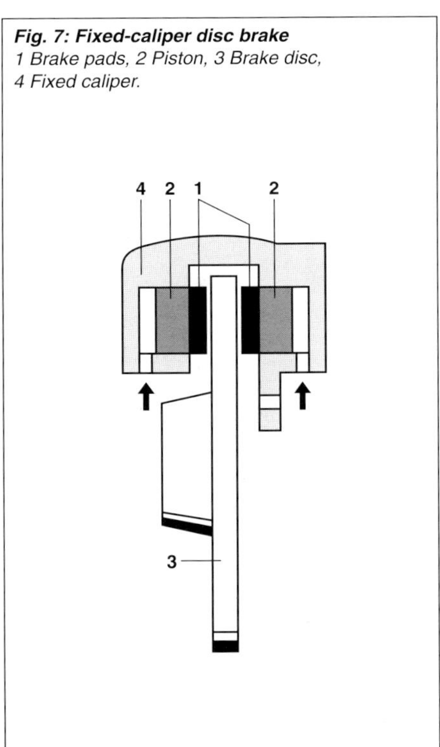

Fig. 7: Fixed-caliper disc brake
*1 Brake pads, 2 Piston, 3 Brake disc,
4 Fixed caliper.*

Floating-caliper disc brakes

Floating-caliper disc brakes (Fig. 8) use only a single piston to press the brake pad against the disc (or rotor). The reactive force shifts the floating-caliper housing, which then presses the opposite pad against the other side of the disc. Due to their more modest installation dimensions, these brakes are required on all vehicles with limited space (negative steering offset[1])). This design is also less sensitive to thermal loads; vapor-bubble formation in the brake fluid, and the resulting brake failure, are extremely rare. Pad replacement is also a model of simplicity: After releasing the attachment bolt, one can tilt the caliper upward to remove the pads. These brakes also adjust themselves in a manner similar to that of the fixed-caliper units.

The fluid in the pressure chamber responds to pedal applications by shifting the piston (2) to the left. The piston impacts against the inner pad (1, right) and presses it to the left, against the disc (3). The caliper housing (4), which is designed to slide on its support (5), reacts by shifting to the right and pushing the outside pad (1, left) against the other side of the disc. Further pressure increases will be reflected in the brake pads being pressed uniformly against the disc.

Drum brakes

Drum brakes on passenger cars generate braking force at the inner surface of a brake drum (inside shoe brakes). As already mentioned, current passenger-car applications are limited to the rear wheels. Numerous versions of the drum brake are in existence; the "Simplex brake" (Fig. 9) is the version most often used in passenger cars.

Drum brakes require periodic adjustment; this can be manual/mechanical or automatic. The wide range of operating requirements for the various drum-brake

---

[1]) Distance of the wheel contact point from the impact point of the extended steering axis of the wheel on the road plane.

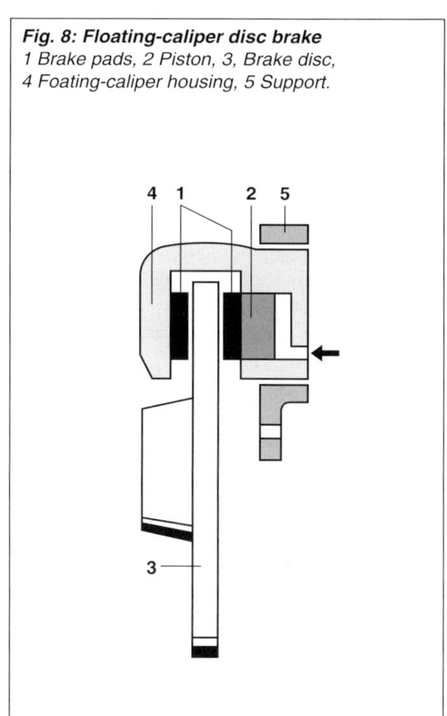

Fig. 8: Floating-caliper disc brake
1 Brake pads, 2 Piston, 3, Brake disc,
4 Foating-caliper housing, 5 Support.

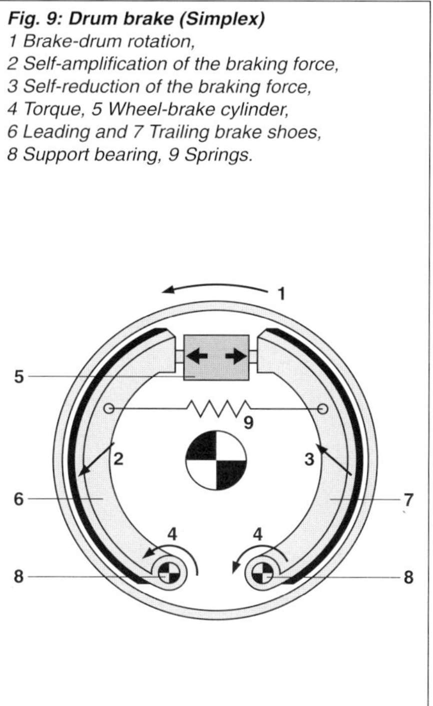

Fig. 9: Drum brake (Simplex)
1 Brake-drum rotation,
2 Self-amplification of the braking force,
3 Self-reduction of the braking force,
4 Torque, 5 Wheel-brake cylinder,
6 Leading and 7 Trailing brake shoes,
8 Support bearing, 9 Springs.

designs is reflected in the large number of adjustment arrangements. Operation of the Simplex brake is controlled by the bilateral pressure exerted by wheel-brake cylinder (5) against the leading (6) and trailing (7) brake shoes to force them up against the brake drum. The leading shoe impacts in the drum's direction of rotation (1), and the trailing shoe (7) in the opposite direction.

On the side opposite the wheel-brake cylinder, the brake drums are located by a support bearing (8) attached to the brake mount. The braking effect is about the same in both directions.

It is a simple matter to equip the Simplex brake with a mechanically controlled parking brake. Springs (9) are employed to retract the shoes.

Automatic adjustment to the desired gap is initiated when the wheel brake is released.

An automatic linkage-rod adjuster adjusts the gap when the friction brakes are released.

Brake linings, pads and discs

The braking (or frictional) force necessary for braking the vehicle is generated at the brake friction linings and the discs (rotors). The coefficient of friction between pads (linings) and disc (drum) plays a decisive role in defining the amount of pedal force required to obtain a given rate of deceleration. This factor is also of major importance in designing brakes for balanced operation, and for vehicle stability when braking.

The exact composition of the pads (preferably asbestos-free) and the coefficients of friction they produce vary greatly from vehicle to vehicle. The vehicle manufacturers devote extensive resources and research to determining the optimum composition for each application. The response characteristics of both pad and disc must remain relatively insensitive to temperatures of up to 700 °C as well as to the influence of contaminants, water, etc.

The brake discs (or rotors) can be divided into two categories: solid and internally-ventilated (Fig. 10). Due to their greater mass, the latter can absorb larger amounts of heat; and at the same time, the air flowing through the internal cooling channels promotes rapid cooling. This is why internally-ventilated rotors are the optimal solution at the front wheels.

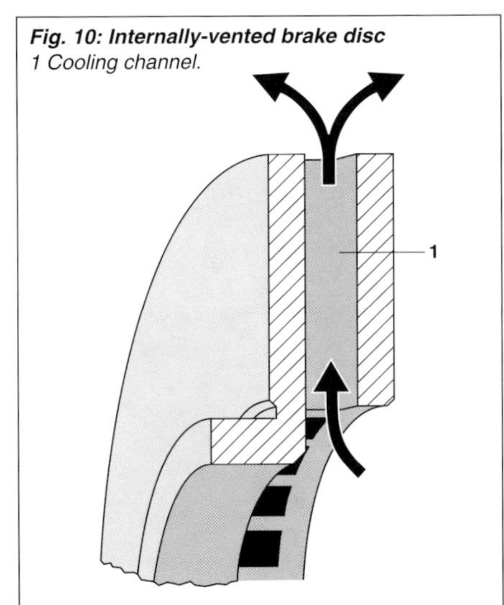

Fig. 10: Internally-vented brake disc
1 Cooling channel.

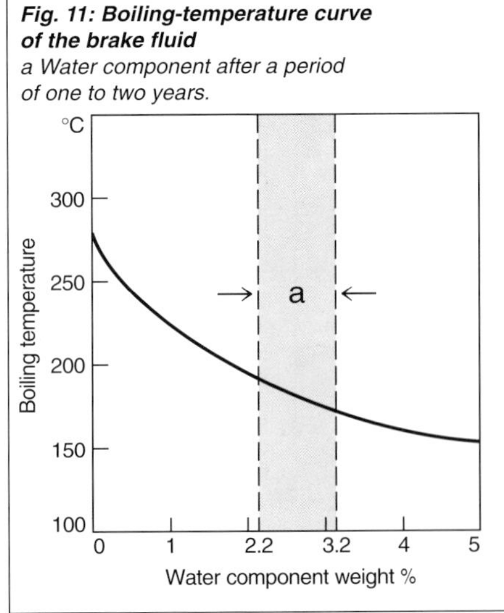

Fig. 11: Boiling-temperature curve of the brake fluid
a Water component after a period of one to two years.

Brake fluids

Brake fluid is the hydraulic medium employed for transmitting force within the brake system. Compliance with stringent requirements is essential in order to guarantee reliable brake-system operation. These requirements are defined in various standards of similar content (SAE J 1703, FMVSS 116, ISO 4925, Table 1).

Requirements

Equilibrium boiling point
The equilibrium boiling point provides an index of the brake fluid's resistance to thermal stress. The heat generated in the wheel-brake cylinders (representing the highest temperatures in the entire braking system) can be especially critical. Vapor bubbles form at temperatures above the brake fluid's momentary boiling point, with brake failure as the ultimate result.

Wet boiling point
The wet boiling point is the equilibrium boiling point of the fluid after it has absorbed moisture under specified conditions. Hygroscopic (glycol-based) fluids are especially sensitive, responding with a sharp drop in boiling point.
Testing of the wet boiling point is directed toward determining the characteristics to be expected from used brake fluid. Brake fluid can absorb moisture through the system's hoses, mostly by diffusion. This is the main reason why the brake fluid should be replaced every 1...2 years. This brake-fluid change is vital for maintaining braking safety; during this operation, particular attention should be devoted to expelling any air trapped in the system (bleeding the brakes).

Viscosity
The viscosity's response to temperature fluctuations should be minimal, as the brake system must provide reliable operation in a wide range of conditions (–40 °C ... +100 °C). Low temperature viscosities are of particular importance in systems with ABS.

Compressibility
The fluid should maintain a low level of compression with minimal temperature sensitivity.

Corrosion protection
FMVSS 116 stipulates that brake fluids will exercise no corrosive effect on those metals which are generally employed in a brake system. The required corrosion protection can only be achieved through the use of additives.

Elastomer swelling
The elastomers in the brake system must be able to adapt to the type of brake fluid being used. Although a small amount of swelling is desirable, it is imperative that it not exceed 16%. Above this figure, the brake fluid tends to weaken the elastomer components. Even minute levels of mineral-oil contamination (mineral-oil-based fluids, solvents) in glycol-based brake fluid can lead to destruction of rubber components (such as seals), with the resulting failure of the brake system.

Chemical composition
Specific chemical structures can be employed to obtain improvements in the above attributes. However, modifications to one characteristic tend to be accompanied by undesired changes in another.

---

[1] 1) FMVSS Federal Motor Vehicle Safety Standard (US safety standards for motor vehicles): DOT Department of Transportation.

Table 1: Brake fluids

Reference standard for testing	FMVSS 116[1]			SAE J 1703
Requirements/Date	DOT3	DOT4	DOT5	11.83
Dry boiling point at least °C	205	230	260	205
Wet boiling point at least °C	140	155	180	140
Cold viscosity at –40 °C mm²/s	1500	1800	900	1800

ABS antilock braking system

Continuing developments in passenger-car brake systems have resulted in powerful and reliable systems capable of furnishing optimum retardation from high rates of speed. Under normal operating conditions these systems can provide fast and effective braking for the vehicle. Braking under more critical conditions, such as those encountered with
- wet or slippery road surfaces,
- driver panic reaction (unanticipated obstacle), and
- mistakes committed by other drivers and pedestrians

can lead to wheel lock when braking. The result: Loss of vehicle steering response as the vehicle loses traction and/or slides from the road.

This is the type of situation in which the antilock braking system ABS comes into play. The system recognizes incipient lock at one or more wheels in time to react by inhibiting further increases or initiating decreases in braking pressure. The result: Vehicle steering response and stability are maintained in a process characterized by optimum braking efficiency.

ABS requirements

ABS must satisfy a wide range of requirements, with particular emphasis on safety considerations regarding dynamic braking response and brake technology:
- The closed-loop brake control system must be capable of maintaining steering response and vehicle stability at all times, regardless of road conditions (extending from dry, high-friction surfaces to glare ice).
- ABS should be capable of exploiting the (coefficient of) friction between tires and road surface to maximum effect, with the proviso that vehicle stability and maintenance of steering response have a higher priority than minimizing stopping distances. These requirements must be fulfilled regardless of whether the driver applies full force immediately, or gradually increases the pedal pressure up to the wheel-lock limit.
- The brake control system must remain operational throughout the vehicle's entire speed range. It must be effective down to walking speed (after which wheel lock is no longer a critical factor in the final distance until the vehicle stops).
- The brake control system must be capable of rapid adaptation to changes in surface traction. For example, on dry roads with occasional ice patches, any wheel lock must be restricted to such brevity that vehicle stability or steering response cannot be impaired. The adhesion on the dry road sections must simultaneously be fully exploited for maximum effectiveness.
- When the brakes are applied on road surfaces affording different traction levels on the two sides of the vehicle (e.g., right tire on ice, left tire on dry asphalt, referred to as "μ-split"), the unavoidable yaw effect (rotational forces centered at the vehicle's vertical axis tending to turn the vehicle sideways) should be slowed to the point where it will be no problem for an average driver to initiate compensatory countersteer.
- During cornering, the vehicle should retain stability and steering response while also braking in the shortest possible distance, the proviso here being that the vehicle's speed remains sufficiently below the cornering limit velocity (the cornering limit velocity is defined as the highest speed at which a vehicle can negotiate a curve of a defined radius without throttle input, and without leaving the defined circle).
- The demands for vehicle stability, steering response and optimal braking also apply on a bumpy road, regardless of braking intensity.
- The closed-loop brake control system must be capable of recognizing and responding to aquaplaning (tires supported by a layer of water between themselves and the road surface). The vehicle's stability and tracking must be maintained.
- Operations to adapt to brake hysteresis (continued braking when the wheel

brake is released) and engine influences (braking with the clutch engaged) must be completed as quickly as possible.
– The vehicle should not respond to oscillations by starting to pitch.
– A circuit must be provided to continually monitor the operating status of the ABS. When this system recognizes a defect or malfunction which could adversely affect braking response, it reacts by switching off the ABS. An indicator lamp alerts the driver to the fact the only the conventional brake system – without ABS – remains available.

Dynamic forces at the braked wheel

Fig. 1 and 2 illustrate the physical relationships that define braking manoeuvres with ABS, whereby the areas in which ABS exercises its closed-loop control function are identified by hatched lines. The patterns (Fig. 1) of curve 1 (dry), 2 (wet) and 4 (black or glare ice) make it clear that shorter braking distances are

achieved with ABS than under panic braking with locked wheels (brake slip $\lambda = 100\%$). In Curve 3 (snow), a wedge of snow increases the braking effect at the locked wheels. Under these circumstances, the main benefits of ABS are in the areas of vehicle stability and steering control.

As the curves for the coefficients of braking force μ_{HF} and lateral forces μ_S in Figure 2 illustrate, the ABS control range must be extended beyond that for brake-slip angle $\alpha = 2°$ when the larger slip angle $\alpha = 10°$ is encountered (that is, with high lateral forces due to the vehicle's high rate of lateral acceleration). When maximum braking force is applied while the vehicle is cornering at high rates of lateral acceleration, ABS reacts by combining rapid active response with (as an example) an initial brake slip of 10%. At $\alpha = 10°$ the initial braking-force coefficient is restricted to $\mu_{HF} = 0.35$, while the lateral-force coefficient with $\mu_S = 0.80$ remains close to its maximum.

As the vehicle continues braking in the curve, the slip rates allowed by ABS rise

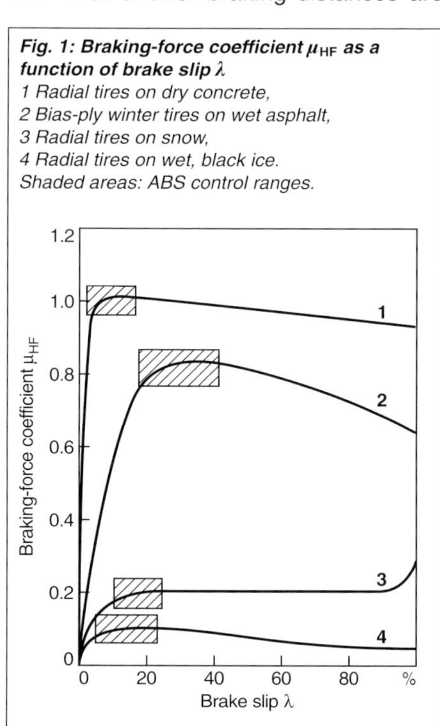

Fig. 1: Braking-force coefficient μ_{HF} as a function of brake slip λ
1 Radial tires on dry concrete,
2 Bias-ply winter tires on wet asphalt,
3 Radial tires on snow,
4 Radial tires on wet, black ice.
Shaded areas: ABS control ranges.

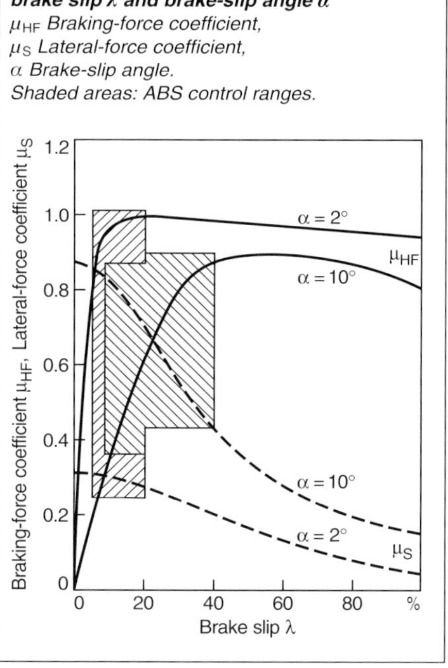

Fig. 2: Braking-force coefficient μ_{HF} and lateral-force coefficient μ_S as a function of brake slip λ and brake-slip angle α
μ_{HF} Braking-force coefficient,
μ_S Lateral-force coefficient,
α Brake-slip angle.
Shaded areas: ABS control ranges.

in inverse proportion to cornering speed and lateral acceleration; the lower lateral-force coefficients that result from the reduction in lateral acceleration are accompanied by higher levels of deceleration. Thus, when the brakes are applied during cornering, the braking forces rise so sharply that the total braking distance is only slightly longer than that achieved under linear braking in comparable conditions.

ABS control loop

The ABS control loop (Fig. 3) consists of the:

Controlled system:
Vehicle with wheel brakes, wheels and friction between tires and road surface.

Disturbance factors:
Road-surface conditions, brake condition, vehicle load and tires (e.g., inadequate tread depth, low tire pressures).

Controller:
Wheel-speed sensors and ABS control unit.

Controlled variables:
Wheel-speed, and the data derived from it on deceleration at the tires' periphery, peripheral wheel acceleration and brake slip.

Reference input variable:
Pressure applied to brake pedal (driver's brake-pressure input).

Manipulated variable:
Braking pressure.

Controlled system

Data processing in the ABS control unit is based on the following, simplified controlled system: A non-driven wheel, one fourth of the vehicle's total mass assigned to this wheel, the wheel brake, and, representing the frictional coupling between tire and road surface, a theoretical curve for friction coefficient vs. slip rate (Fig. 4). This curve is divided into a stable range, characterized by linear rises, and an instable section with a constant response line (μ_{HFmax}). Yet another simplified process is a braking procedure under linear (straight-ahead) operation, corresponding to panic braking.

Figure 5 illustrates the relationships between braking torque M_B (the force which the brake can apply through the tire), or road frictional torque M_R (the force that returns to the wheel from the road/tire frictional coupling), and time t. Also shown is the relationship between peripheral wheel deceleration $(-a)$ and time t. The braking torque displays a linear increase over time. The road-surface frictional torque follows the braking torque with a slight time delay T for as

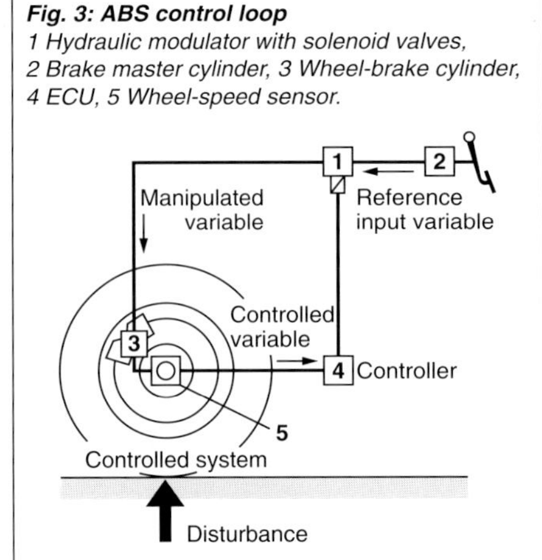

Fig. 3: ABS control loop
1 Hydraulic modulator with solenoid valves,
2 Brake master cylinder, 3 Wheel-brake cylinder,
4 ECU, 5 Wheel-speed sensor.

Fig. 4: Idealized braking-force coefficient/slip curve
a) Stable area, b) Unstable area.
λ_K Optimum brake slip,
μ_{HFmax} Maximum braking-force coefficient.

long as the braking process remains within the stable range on the braking-force coefficient/slip curve. The curve reaches its maximum level (μ_{HFmax}) after approximately 130 ms; at this point it enters the instable range on the braking-force coefficient/slip curve.

The braking torque M_B continues to increase, while the road frictional torque M_R cannot rise any further but remains constant; as indicated by the curve for braking-force coefficient vs. slip. In the period between 130 and 240 ms (at which the wheel locks), the differential $M_B - M_R$, which remains minimal in the stable range, quickly assumes much larger proportions. This torque differential provides a precise index of peripheral wheel deceleration ($-a$) at the braked wheel (Fig. 5, below). The peripheral deceleration, restricted to low levels in the stable range, increases rapidly after the transition to the instable range. The result is an inversion in the response patterns for the stable and instable ranges on the braking-force coeffient/slip curve. ABS exploits this opposed response.

Controlled variables

Selection of suitable controlled variables is a major factor in determining the efficiency of ABS control. The basis is provided by the signals from the wheel-speed sensors, which the control unit (ECU) employs to calculate the wheel's peripheral deceleration and acceleration, brake slip, reference speed and vehicle deceleration. On their own, neither the peripheral wheel deceleration (acceleration) nor the brake slip is suitable for use as a controlled variable, as the reaction that a driven wheel displays to braking will differ vastly from that of a non-driven wheel. However, these variables can be used to obtain satisfactory results when combined in defined logical relationships. Brake slip cannot be measured directly. The ECU therefore calculates a representative figure. This is based on the reference speed that corresponds to the characteristic velocity for optimal braking conditions (optimum brake slip). The ECU determines this based on the con-

stant flow of wheel-speed signals which it receives from the wheel-speed sensors. It selects a "diagonal" (for instance, right front and left rear wheel) and uses this as the basis for a reference speed. Under moderate braking, the reference speed will usually be based on the diagonal wheel which is turning faster. During panic stops with active ABS control, the wheel speeds diverge from vehicle speed, and unless a correction factor is used, they are unsuitable for calculating the reference speed. During the control phase, the ECU generates this speed based on a ramp-shaped extrapolation of the speed at the start of the cycle. The unit processes logical signals and evaluates relationships to calculate the precise slope angle of the ramp.

Ideal closed-loop control of the braking process is available when the data on brake slip and on acceleration/ deceleration rates at the wheel's periphery are supplemented by secondary data for vehicle deceleration, which can be employed to adjust the logical processes in the ECU. This concept has been implemented in the Bosch antilock braking system (ABS).

Fig. 5: Initial braking process, simplified
($-a$) Peripheral wheel deceleration,
($-a_{max}$) Maximum peripheral wheel deceleration,
M_B Braking torque, M_R Road frictional torque,
M_{Rmax} Maximum road frictional torque,
T Time delay.

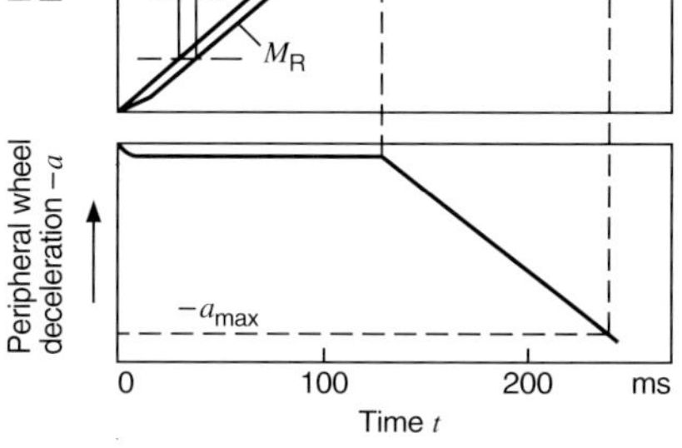

Controlled variables for non-driven wheels

The acceleration and deceleration rates at the circumference of the wheel can generally serve as reliable controlled variables for non-driven wheels, as well as for driven wheels provided that the driver disengages the clutch while braking. The reason has to do with the opposed response patterns displayed by the controlled system in the stable and instable ranges of the braking-force coefficient/slip curve:

The rate of peripheral deceleration available in the stable range is limited – in other words when the driver applies more force to the brake pedal, the vehicle responds with increased deceleration, without the wheels locking.

The instable range displays a different pattern. Here, a minimal increase in pedal pressure will be enough to induce immediate wheel lock. This response pattern can often be employed in using the wheel's peripheral deceleration and acceleration rates to determine the slip rate corresponding to optimal braking.

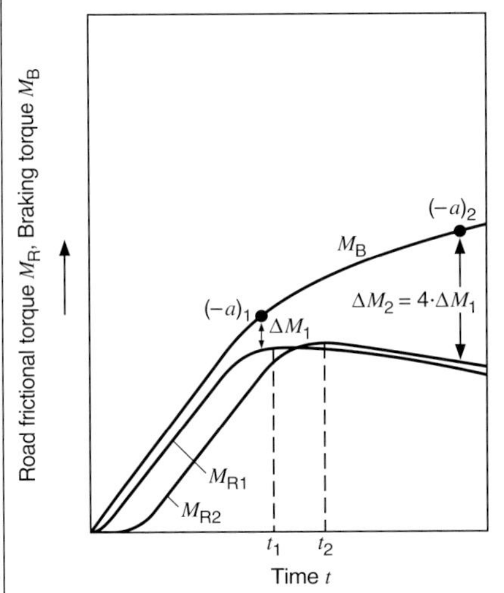

Fig. 6: Initial braking process for a non-driven wheel and a driven wheel coupled with the engine
Index 1: Non-driven wheel,
Index 2: Driven wheel.
$(-a)$ Deceleration theshold,
ΔM Torque difference $M_B - M_R$, simplified.

Any fixed threshold defining the peripheral deceleration beyond which ABS assumes active control should not exceed the maximum potential vehicle deceleration by any substantial amount. This is especially important when light initial brake-pedal applications are followed by increasing hard braking. An excessively high threshold would allow the wheels to shift well into the instable range on the braking-force coefficient/slip curve before ABS would re-spond to the incipient instability. When a wheel first reaches the fixed threshold for peripheral deceleration during a panic stop, the system should not respond by automatically reducing the brake pressure at this wheel. If modern tires are being used on a high-traction surface, this could lead to a sacrifice in valuable brak-ing distance; this factor is of particular importance in braking initiated from high speeds.

Controlled variables for driven wheels

If the vehicle is braked with first or second gear engaged, the engine will act on the driven wheels. This will be reflected in a substantial increase in the wheels' mass moment of inertia Θ_R – the wheels will react as though they are much heavier than they really are. The ultimate result is a proportional loss in the degree to which the wheels' peripheral deceleration rates react to braking-torque variations within the instable range of the braking-force coefficient/slip curve.

The non-driven wheels display sharply opposed response patterns depending upon whether they are in the stable or instable range of the braking-force coefficient/brake-slip curve. The effect described above negates these differences to such a degree that the wheel's peripheral deceleration (as controlled variable) frequently ceases to furnish an adequate index for determining the braking slip with the greatest possible friction. It becomes necessary to consult a supplementary variable, similar to the brake slip; this can then be applied in combination with the peripheral deceleration rate.

Figure 6 compares a braking process for a braked non-driven wheel and for a braked driven wheel which remains coupled to the engine. In this example the engine inertia quadruples the effective mass moment of inertia at the wheel. The non-driven wheel exceeds a defined peripheral-deceleration threshold $(-a)_1$ soon after leaving the stable range on the braking-force coefficient/brake-slip curve. Because the moment of inertia at the driven wheel is multiplied by a factor of 4, four times the differential in inertial moment

$$\Delta M_2 = 4 \cdot \Delta M_1$$

must be achieved before the threshold $(-a)_2$ is exceeded. At this point the driven wheel could already be well into the instable range of the braking-force coefficient/slip curve, with commensurate sacrifices in vehicle stability as the result.

Typical control cycles

Closed-loop braking control on high-traction surfaces (high braking-force coefficients)

When ABS closed-loop control of the braking process is triggered on a high-traction road surface (surface with a high coefficient of friction), in order to avoid suspension and drivetrain resonances, the subsequent pressure rise must be protracted (slowed by a factor of 5...10) compared to the initial braking phase. The curves in Figure 7 represent this state in which braking control operates under conditions of high braking-force coefficient.

The wheel's peripheral deceleration moves beyond the defined threshold $(-a)$ at the end of Phase 1, and the solenoid valve shifts to its "maintain pressure" position. It is still too early to start reducing the brake pressure, as the threshold $(-a)$ could be exceeded within the stable range on the braking-force coefficient/ slip curve, and valuable braking distance would be "sacrificed." The reference

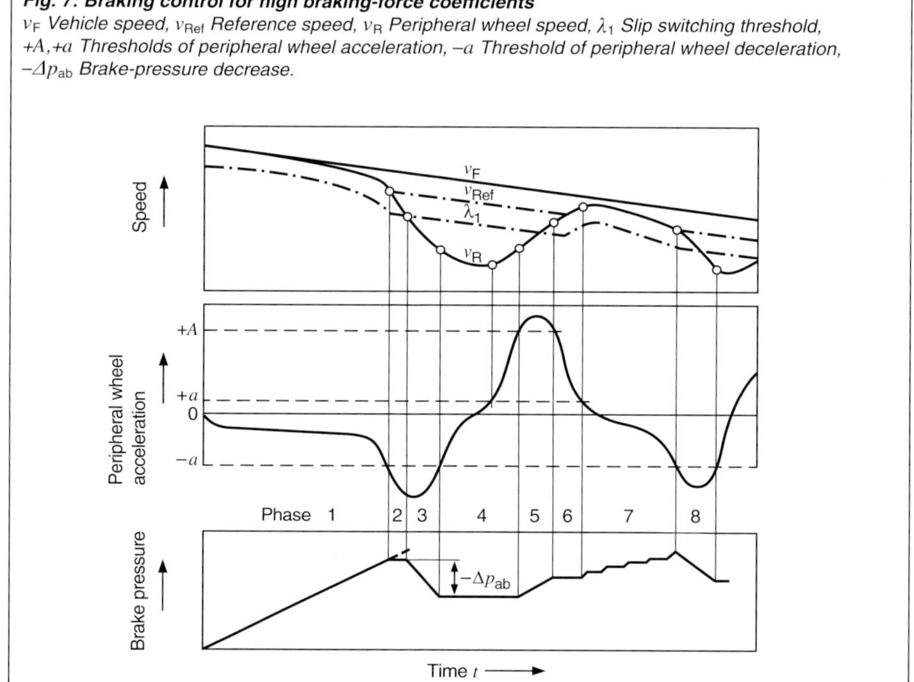

Fig. 7: Braking control for high braking-force coefficients
v_F Vehicle speed, v_{Ref} Reference speed, v_R Peripheral wheel speed, λ_1 Slip switching threshold,
$+A, +a$ Thresholds of peripheral wheel acceleration, $-a$ Threshold of peripheral wheel deceleration,
$-\Delta p_{ab}$ Brake-pressure decrease.

speed v_{Ref} is reduced at the same time in accordance with a defined ramp. The reference speed serves as the basis for determining the slip switching threshold λ_1. At the end of phase 2, the wheel's peripheral velocity v_R drops below the λ_1 threshold. The solenoid valve reacts by shifting to its "pressure release" position. The brake pressure then continues to drop until the wheel's peripheral deceleration exceeds the threshold $(-a)$ again. At the end of the 3rd phase it falls back below the threshold $(-a)$; this is followed by a pressure-holding phase of defined duration. During this phase, the wheel's peripheral acceleration has increased enough to exceed the threshold $(+a)$. The pressure remains constant.

At the end of Phase 4 the wheel's peripheral acceleration exceeds the relatively pronounced threshold $(+A)$. The brake pressure then continues to increase for as long as the acceleration remains above the threshold $(+A)$.

In Phase 6 constant pressure is maintained in response to the fact that the threshold $(+a)$ has been exceeded. This state indicates that the wheel has entered the stable range on the braking-coefficient/slip curve, and is slightly underbraked.

The brake pressure is now built up in stages (Phase 7) in a process that continues until the wheel's peripheral deceleration again exceeds the threshold $(-a)$ (end of Phase 7). This time the brake pressure is reduced immediately, without generation of a λ_1 signal.

Closed-loop braking control on low-grip (slippery) road surfaces (low braking-force coefficients)

In contrast to the behavior patterns on high-grip road surfaces, on slippery roads even light pressure on the brake pedal often suffices to induce wheel lock. The wheels also need substantially more time to emerge from a high-slip period and to accelerate again.

The logical circuits in the ECU recognize the current road conditions and adapt the ABS reponse characteristics accordingly.

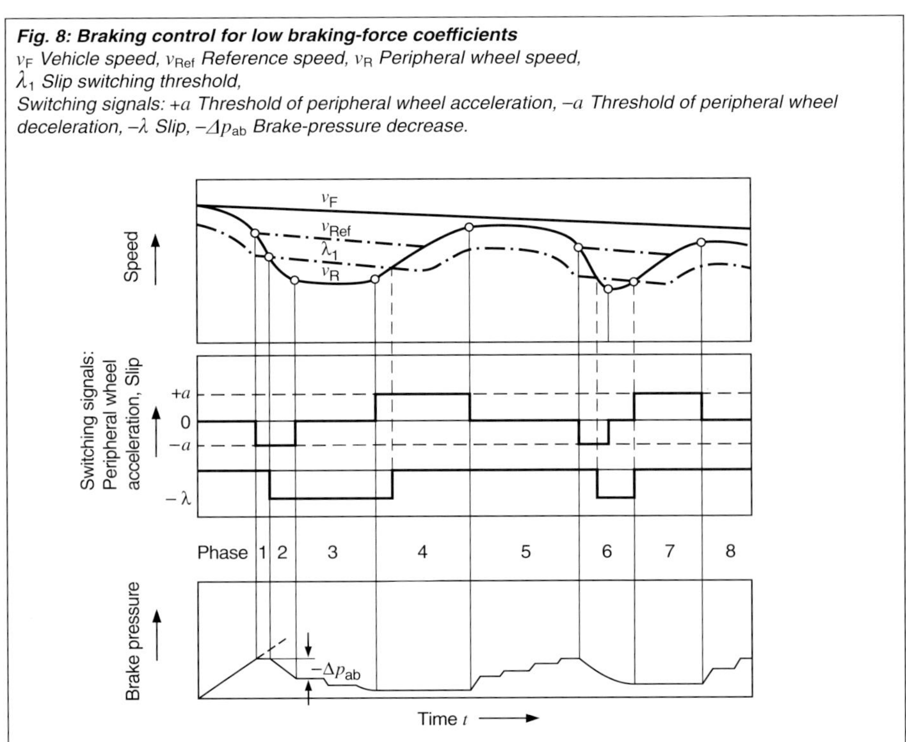

Fig. 8: Braking control for low braking-force coefficients
v_F Vehicle speed, v_{Ref} Reference speed, v_R Peripheral wheel speed,
λ_1 Slip switching threshold,
Switching signals: $+a$ Threshold of peripheral wheel acceleration, $-a$ Threshold of peripheral wheel deceleration, $-\lambda$ Slip, $-\Delta p_{ab}$ Brake-pressure decrease.

Figure 8 illustrates a typical braking-control pattern for low braking-force coefficients.

In Phases 1 and 2 the braking-control process is the same as that applied on high-traction (high-grip) surfaces.

Phase 3 starts with a brief pressure-holding period, followed by a very rapid comparison between wheel speed and the slip switching threshold λ_1. The wheel's peripheral speed is lower than the value for the slip switching threshold, so braking pressure is reduced for a short defined period.

The next step is a second short pressure-holding phase. The system then again compares peripheral speed and slip switching threshold λ_1, which results in a short, defined pressure-release period. In the subsequent pressure-holding phase, the wheel again accelerates to the point where its peripheral acceleration exceeds the threshold ($+a$). This initiates a renewed pressure-holding period, extending until the peripheral speed drops back below the threshold again ($+a$) (end of Phase 4). Phase 5 is characterized by the graduated pressure increase familiar from the previous section. Finally, in Phase 6, pressure is released to initiate a new control cycle.

In the cycle described above, the control logic recognizes that two supplementary pressure-reduction operations are required in order to re-accelerate the wheel following the pressure reduction initiated by the signal ($-a$). The wheel remains in the high-slip range for a relatively extended period, with negative effects for vehicle stability and steering control.

To improve both these factors, the system continuously monitors and compares the wheel's peripheral speed and the slip switching threshold λ_1 in this and in the following control cycles.

Consequently, Phase 6 is characterized by continuous reductions in brake pressure; which are maintained until the peripheral speed exceeds the threshold ($+a$) in Phase 7. Due to the continuous pressure reductions, the wheel spends only a minimal amount of time in the high-slip range; the ultimate result is en-hanced stability and steering response relative to those obtained in the first control cycle.

Closed-loop braking control with yawing-moment buildup delay

When the brakes are applied on a road with different surface conditions on the lane's crown and on its shoulder – e.g., with the left wheels on dry asphalt, and the right tires running on ice – the result is a severe difference in the braking forces at the front wheels. This induces a rotational force centering around the vehicle's vertical axis – the yawing moment (Fig. 9).

Heavier passenger cars with an extended wheelbase display a relatively high vehicular moment of inertia around the vertical axis. On these vehicles, the transition to yaw is relatively slow, and the driver will have enough time to initiate effective steering corrections during ABS braking. Smaller vehicles with a short wheelbase and a low moment of inertia present a different picture. Here, the ABS

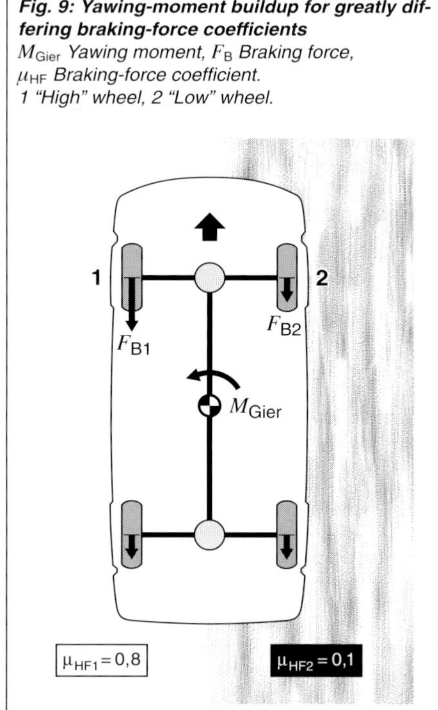

Fig. 9: Yawing-moment buildup for greatly differing braking-force coefficients
M_{Gier} Yawing moment, F_B Braking force, μ_{HF} Braking-force coefficient.
1 "High" wheel, 2 "Low" wheel.

needs to be supplemented by a yawing-moment buildup delay system (GMA) to ensure that these vehicles remain stable and responsive to driver inputs under panic braking on lanes affording different traction levels at crown and shoulder. This yawing-moment buildup delay slows down pressure buildup in the wheel-brake cylinder at the front wheel displaying the higher braking-force coefficient (the "high" wheel).

Fig. 10 illustrates the concept behind the yawing-moment buildup delay system: Curve 1 shows the brake master-cylinder pressure p_{HZ}. Without GMA, the wheel on asphalt attains the pressure p_{high} (Curve 2), and the wheel on ice the pressure p_{low} (Curve 5) within a brief period; each wheel brakes to provide the maximum available deceleration (individual control).

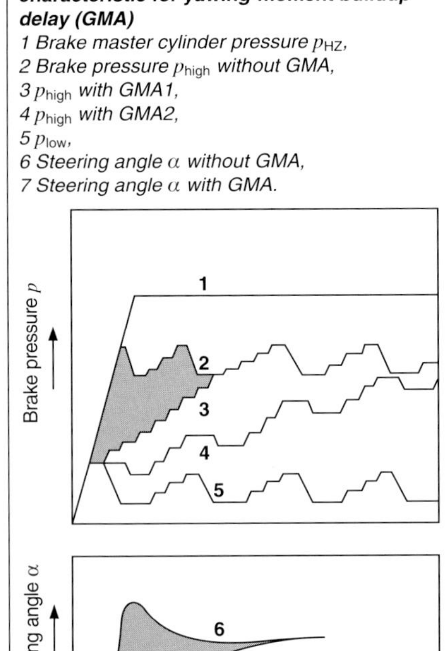

Bild 10: Brake-pressure/steering-angle characteristic for yawing-moment buildup delay (GMA)
1 Brake master cylinder pressure p_{HZ},
2 Brake pressure p_{high} without GMA,
3 p_{high} with GMA1,
4 p_{high} with GMA2,
5 p_{low},
6 Steering angle α without GMA,
7 Steering angle α with GMA.

GMA 1

The GMA 1 system is employed on vehicles with relatively uncritical response patterns. In the initial braking phase, this system increases the brake pressure at the "high" wheel (Curve 3) in defined increments, as soon as pressure is released at the "low" wheel for the first time as a result of incipient lock. Once the "high" wheel reaches the brake pressure corresponding to its lock level, the signals from the "low" wheel cease to be relevant and it is individually controlled so that it can provide maximum braking efficiency. This procedure ensures that these types of vehicle retain satisfactory steering response characteristics during panic braking on heterogeneous road surfaces. Because maximum braking pressure with only a minute delay (750 ms) is applied to the "high" wheel, the resulting increase in braking distance relative to that of a vehicle without GMA is minimal.

GMA 2

GMA 2 is installed in vehicles with more critical response characteristics. As soon as the brake pressure is released at the "low" wheel, the system commands the ABS solenoid valve at the "high" wheel to start defined pressure-holding and reduction cycles (Fig. 10, Curve 4). The renewed pressure increase at the "low" wheel then triggers a graduated pressure rise at the "high" wheel, whereby the pressure accumulation times on the high side are longer by a specified factor than those at the low wheel. This pressure-metering process is not restricted to the intial control cycle. Instead, it is maintained for the duration of the brake application.

The effects of yaw on steering response become more critical the higher the vehicle speed at the time braking starts. GMA 2 defines vehicle speed by dividing it into four categories, with different levels of yawing-moment buildup delay for each. The high-speed ranges are characterized by progressively shorter pressure-accumulation periods at the "high" wheel, while this period increases continually at the "low" wheel. This is intended to reduce the yawing-moment build-up particularly at high vehicle speeds. The lower section of Figure 10 illustrates two steering-angle progressions under braking, with GMA (Curve 6) and without it (Curve 7).

An ideal system for inhibiting yaw is a compromise between steering control and acceptably short braking distances. Bosch designs the system for each individual vehicle in cooperation and consultation with the respective manufacturer.

Another important aspect for GMA applications is the vehicle's behavior when braking during cornering. GMA responds to high-speed braking in curves by increasing the dynamic loads on the front suspension while decreasing them at the rear. The result is higher levels of lateral force at the front wheels accompanied by reductions at the rear. The torque forces are thus directed toward the inside of the curve; the vehicle starts to slide toward the inside, and steering corrections are difficult to apply (Fig. 11, top).

To prevent this situation from arising, GMA is also equipped with a supplementary lateral-acceleration switch designed to deactivate this function at rates of lateral acceleration exceeding 0.4 g. As a result, braking is accompanied by a high level of braking force at the outside front wheel, generating torque directed toward the outside of the curve. This rotational force compensates for the inward-directed torque generated by the lateral forces; the vehicle assumes a mildly understeering attitude to provide the driver with good control (Fig. 11, bottom).

GMA 1 is implemented using an ECU featuring large-scale integrated circuitry. GMA 2 employs two additional microprocessors designed for parallel operation with mutual monitoring.

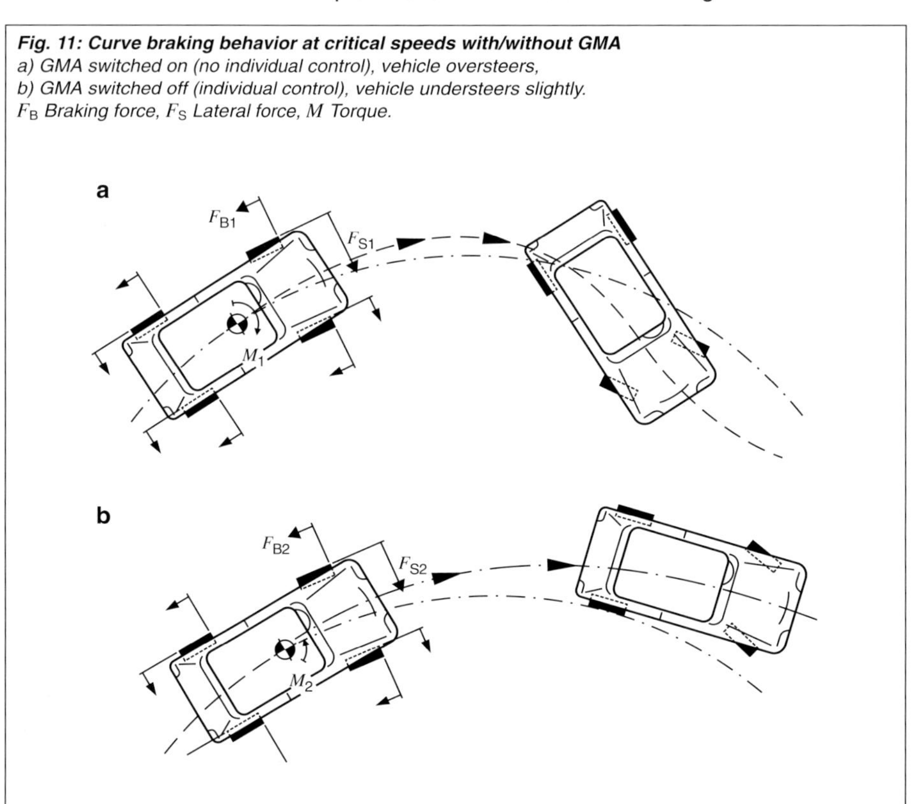

Fig. 11: Curve braking behavior at critical speeds with/without GMA
a) GMA switched on (no individual control), vehicle oversteers,
b) GMA switched off (individual control), vehicle understeers slightly.
F_B Braking force, F_S Lateral force, M Torque.

Closed-loop braking control for AWD
The most important criteria in assessing the various all-wheel-drive systems (Figure 12) are traction, vehicle dynamics and braking response.

Once the differential locks are engaged, the conditions under which ABS must operate change, and supplementary ABS features are required.

Locking the rear differential produces a continuous positive coupling between the rear wheels. They rotate at the same speed, and with regard to braking torque (at the two wheels) and both surface coefficients of friction (between the two tires and the respective road-surface contact patches), they react like a solid body. The "select-low" mode is cancelled at the rear axle, and both rear wheels exploit their braking forces to maximum effect. As soon as the inter-axle lock is engaged, the system responds by forcing the front and rear wheels to assume the same mean rotational speed. At this point a dynamic connection exists between all of the wheels, and the engine drag torque (engine braking under trailing throttle) and engine inertia affect the response at all wheels.

All-Wheel-Drive (AWD) systems
In order to ensure that the ABS continues to operate at optimum efficiency under these special conditions, supplementary functions are required; these vary according to individual AWD configuration (Fig. 12):

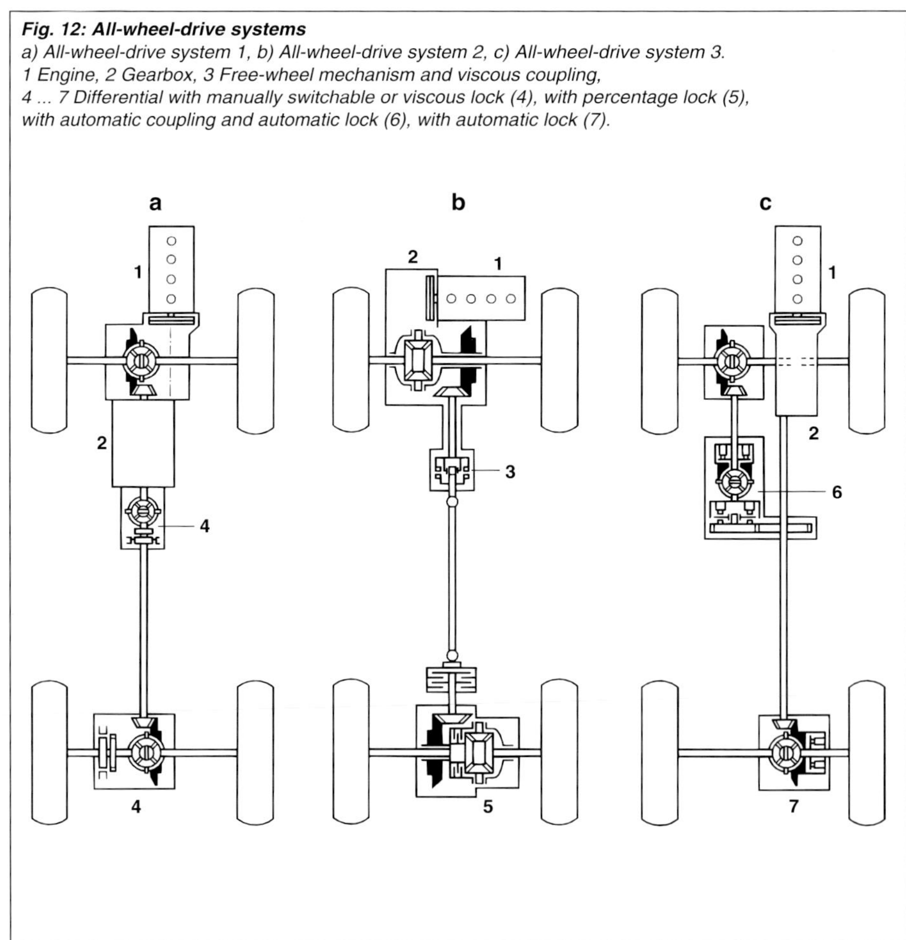
Fig. 12: All-wheel-drive systems
a) All-wheel-drive system 1, b) All-wheel-drive system 2, c) All-wheel-drive system 3.
1 Engine, 2 Gearbox, 3 Free-wheel mechanism and viscous coupling,
4 ... 7 Differential with manually switchable or viscous lock (4), with percentage lock (5),
with automatic coupling and automatic lock (6), with automatic lock (7).

All-Wheel-Drive system 1

On category 1 all-wheel-drive systems – featuring manually-activated locks or permanent (viscous-coupling) interaxle locks – a positive coupling exists between the rear wheels, which also maintain the same mean rotational speed as the front wheels. As explained above, the rear-axle lock cancels the "select low" mode at the rear wheels, meaning that maximum braking force can be effectively applied at each of these wheels. When the brakes are applied on heterogeneous surfaces, the difference in braking force at the rear axle generates a yaw force capable of inducing critical changes in vehicle stability. The vehicle would become uncontrollable if the maximum braking force differential were generated rapidly at the front wheels at the same time. This type of 4-wheel drive makes it necessary to provide GMA at the front axle in order to ensure that the left and right wheels continue to provide stable operating and steering control even when the surfaces supporting them are characterized by starkly different traction conditions.

On all-wheel drive vehicles, the engine drag torque, which affects all the wheels, must be reduced if ABS is to continue to operate efficiently on low-traction surfaces. This is accomplished by increasing the idle speed or by incorporating an engine drag torque control capable of applying just enough throttle to prevent excess engine braking. The reduction in the wheels' sensitivity to variations in traction coefficients on low-adhesion surfaces resulting from the effects of engine inertia, must also be compensated for by incorporating additional features in the brake control system. The ECU must incorporate supplementary signal-processing and logical functions to compensate for this dynamic coupling between the engine's inertial mass and all of the vehicle's wheels. A linear-deceleration switch allows the system to recognize low-traction surfaces with μ_{HF} lower than 0.3. For the purposes of evaluating brake reponse on this type of road, the response threshold ($-a$) for peripheral dece-leration is halved. The rise in reference speed, progressing at a lower rate, is restricted to defined minimal levels. Incipient wheel lock is recognized at an early point, and "sensitive" control is maintained. Abrupt throttle applications on a slippery road surface can cause all the 4WD vehicle's wheels to spin at once. The signal-processing circuitry responds to this situation by ensuring that increases in reference speed based on the slipping wheels do not exceed the maximum potential vehicular acceleration. During subsequent braking, the initial ABS pressure release process is triggered by a signal ($-a$) and a defined minimal difference in wheel speed.

All-Wheel-Drive system 2

Wheelspin at all four wheels is also possible with System 2 (viscous one-way interaxle coupling, proportional rear axle lock); thus the supplementary signal processing employed with System 1 is also required here.

No other supplementary measures are required for satisfactory ABS operation, as the one-way coupling disconnects the wheels under braking. However, a provision for engine drag-torque control can improve the system's operating characteristics.

All-Wheel-Drive system 3

As with the other two systems, special signal-processing arrangements are needed to cope with slippage at all four wheels on all-wheel-drive system 3 (automatic lock activation). An additional factor is that the locks release automatically each time the brakes are applied. No supplementary measures are required for satisfactory ABS operation.

ABS Versions

Version ABS 2S

The first antilock braking system, ABS 2S (Fig. 13), entered series production in 1978, when digital electronics had advanced to a point at which it became possible to monitor complex braking processes and to react in fractions of a second. This extremely flexible system can be integrated within the basic conventional brake system without major modifications. It operates as follows (Fig. 14):

During vehicle operation, the system's sensors monitor wheel speed at both front wheels, and at either the differential or at both rear wheels. When, based on the sensor signals, the ECU recognizes impending lock, it responds by actuating the solenoid valves for the affected wheels in the hydraulic modulator. Individual solenoid valves assigned to each front wheel control the wheels' response to ensure that each tire makes its maximum potential contribution to effective braking – regardless of conditions at the other wheels (individual closed-loop control). At the rear axle, the wheel with the lower braking-force coefficient determines the pressure to be applied at both wheel brakes (select-low principle). When the front wheel/rear wheel configuration is used, a single solenoid valve controls the rear wheels. When the diagonal configuration is used, two solenoid valves are required for this purpose. There is a slight increase in braking distances with this control principle, but this is more than outweighed by the enhanced stability which it provides. The ECU shifts the solenoid valves between three different positions:

– The first (non-energized) position connects the brake master cylinder and wheel-brake cylinder, braking-pressure buildup can take place.

– The second position (energization with 50% of maximum current) separates the wheel-brake cylinder from the brake master cylinder and from the return line, thus holding wheel-brake pressure at a constant level.

– The third position (energization with maximum current) isolates the brake master cylinder while at the same time connecting wheel brake and return (or discharge) line, allowing the wheel brake pressure to drop. This configuration permits not only continuous, but also stepped (and therefore moderate), increases and reductions in pressure.

Depending upon road-surface conditions, the number of control cycles per second varies between 4 and 10. ABS achieves this rapid control by means of electronic signal processing and fast response times.

Each time the vehicle is started, and at the end of each trip, the ECU initiates a test cycle to verify the status of the

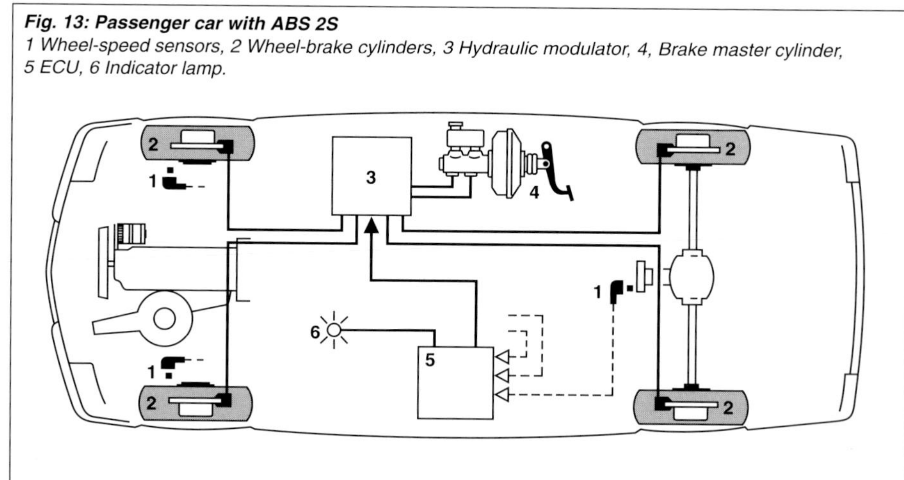

Fig. 13: Passenger car with ABS 2S
1 Wheel-speed sensors, 2 Wheel-brake cylinders, 3 Hydraulic modulator, 4, Brake master cylinder, 5 ECU, 6 Indicator lamp.

Fig. 14: Brake-pressure modulation
a) Pressure buildup, b) Hold pressure, c) Reduce pressure.
1 Wheel-speed sensor, 2 Wheel-brake cylinder, 3 Hydraulic pressure modulator, 3a Solenoid valve,
3b Accumulator, 3c Return pump, 4 Brake master cylinder, 5 ECU.

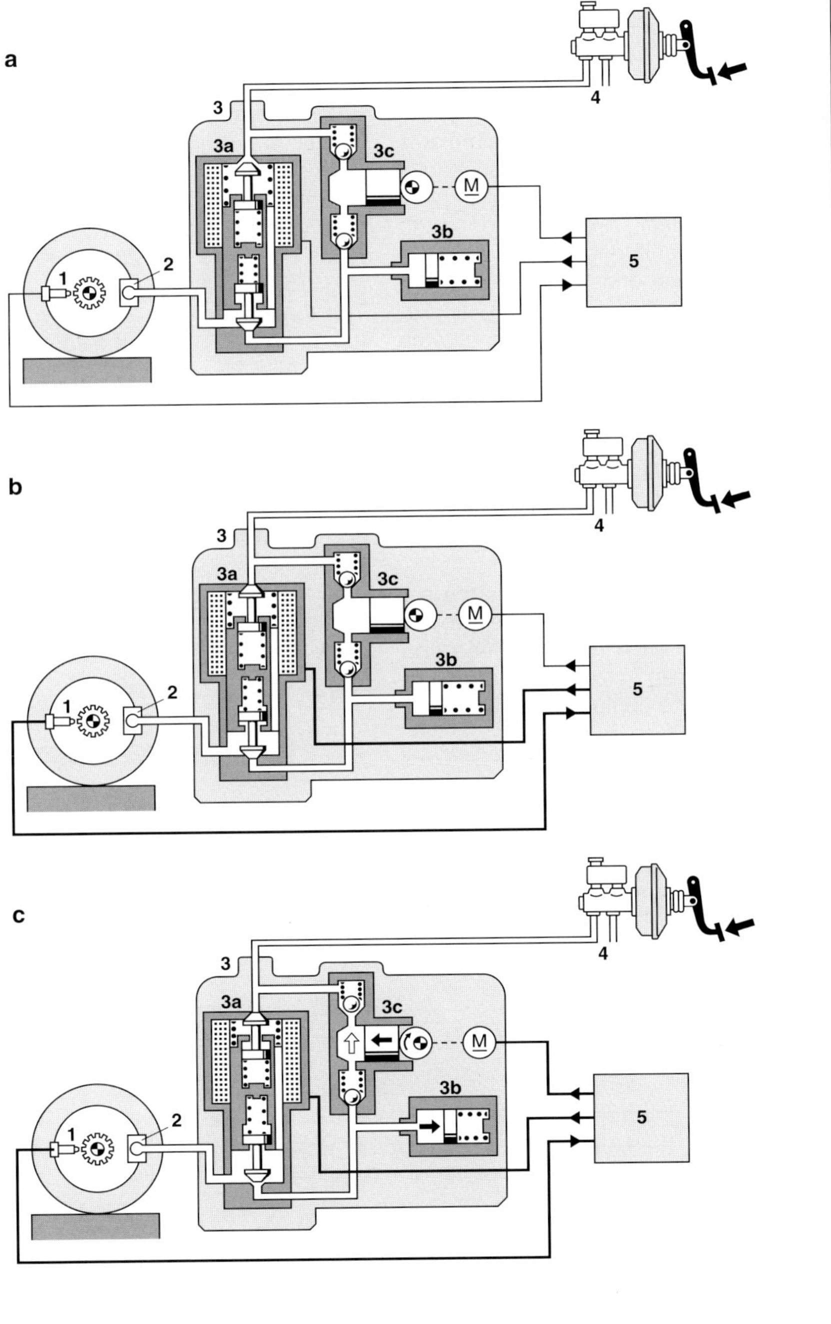

controller, the safety circuits and all peripheral systems.

This test cycle makes it possible to check the status of circuits that remain inactive when ABS control has not been triggered, but which could have an effect on the brake-control process in an emergency.

One important component within this test cycle is the self-test for the monitoring circuit. This procedure simulates errors in order to analyse the response of the circuits, including the deactivation paths. This test prevents "dormant" errors in the monitoring circuits.

The system responds to a recognized fault by switching off the ABS, a process accompanied by a warning lamp lighting up to alert the driver to the fact that only the basic, conventional brake system is still available. This safety circuit, complete with computer-controlled error simulation, guarantees that high security standards are maintained.

Version ABS 5.0

The ABS 5.0 antilock braking system is a more advanced version of the proven ABS 2S unit. FMEA (Failure Mode and Effects Analysis) processes are employed to ensure comprehensive system integrity. ABS 5.0 is distinguished by the following salient characteristics:

– Modular construction for flexibility in application,
– Return principle with closed brake circuits, and
– Computer redundancy (dual-processor principle) with comprehensive monitoring software.

The chief distinction via-à-vis ABS 2S is represented by the solenoid valves in the hydraulic modulator unit. ABS 2S operates with 3/3 solenoid valves, but ABS 5.0 is equipped with 2/2 solenoid valves (Fig. 15).

The braking-force distribution patterns used with ABS 5.0 are the same as those of the ABS 2S:

For II patterns (braking-force distribution with separate circuits for front and rear wheels: Each circuit brakes a different axle), both a 3 and a 4-channel version are available.

For X patterns (diagonal braking-force distribution: Each circuit brakes one wheel at the front axle, and the diagonally-opposed wheel at the rear axle), there is a 4-channel version available (Fig 16).

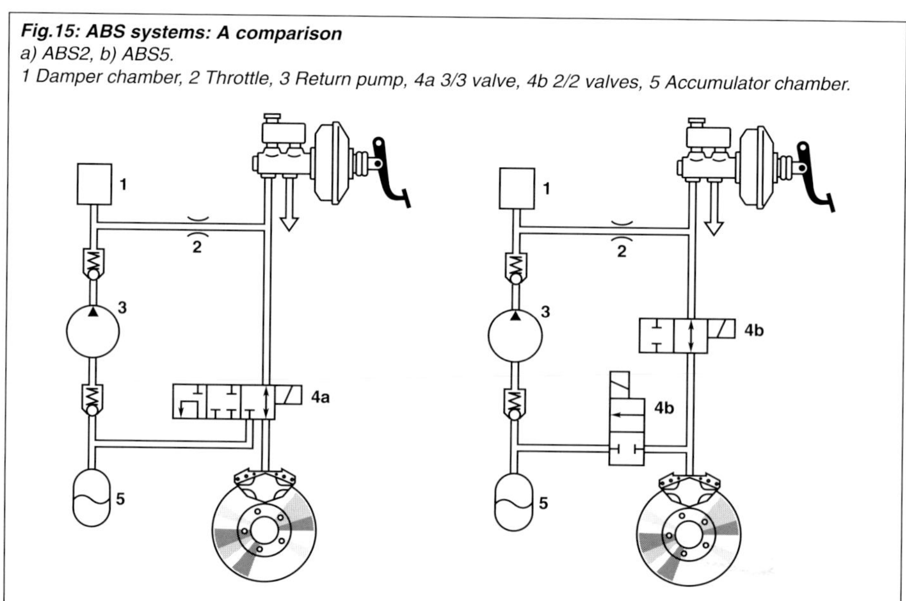

Fig.15: ABS systems: A comparison
a) ABS2, b) ABS5.
1 Damper chamber, 2 Throttle, 3 Return pump, 4a 3/3 valve, 4b 2/2 valves, 5 Accumulator chamber.

Closed-loop control process

The signals transmitted by the wheel-speed sensors from the individual wheels are used by each microcontroller as the basis for calculating all control data. The calculated wheel speeds are employed to derive a "semi-proportional" definition of wheel slip as a function of the vehicle's reference speed (see section on "Typical control cycles"). Under ideal conditions the reference speed will be found in the vicinity of the slip rate with maximum friction. In standard operation the speed of the fastest tire will generally be selected to provide an auxiliary reference speed. This data, filtered to compensate for time factors, provides an index of the vehicle's rate of deceleration.

Monitoring functions

The ECU employs two redundant (parallel with mutual checks) microcontrollers to monitor the status of the entire electronic signal-processing apparatus, the logic circuits and the monitoring software. The two microcontrollers should respond to identical input signals by generating identical output signals. The ABS responds to extended logical variations between control signal and feed-back with error recognition followed by system switch off. This feature guarantees the 100% monitoring of the signal-processing and logical functions.

All lines to the following peripheral components are monitored continuously:
- sensors,
- solenoid valves, and
- brake-light switch.

The ECU also monitors:
- the operational status of the return-pump motor,
- pump-motor inertial rotation in the return pump, based on voltage measurements,
- voltage level in order to detect open circuits,
- both wheel and reference speeds during initial acceleration,
- static slip during normal vehicle operation (e.g., due to variations in wheel diameter),
- dynamic rotation speed at high speeds (to inspect for open lines),
- triggering times for solenoid valves, and
- monitoring errors traceable to interference from extraneous error sources.

Each time the vehicle is started, brief

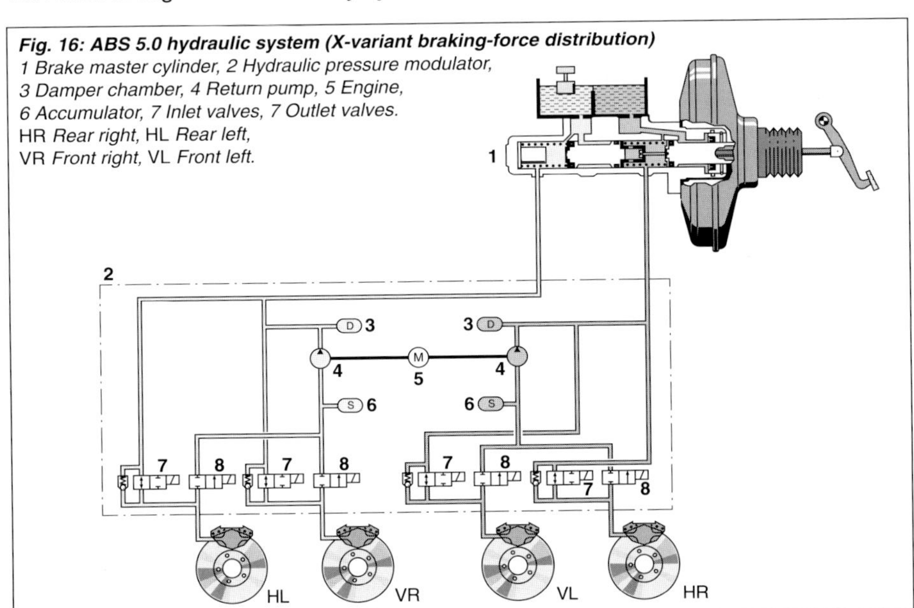

Fig. 16: ABS 5.0 hydraulic system (X-variant braking-force distribution)
1 Brake master cylinder, 2 Hydraulic pressure modulator,
3 Damper chamber, 4 Return pump, 5 Engine,
6 Accumulator, 7 Inlet valves, 7 Outlet valves.
HR Rear right, HL Rear left,
VR Front right, VL Front left.

electrical currents are transmitted to the solenoid valves and the pump motor as soon as the vehicle speed exceeds 6 km/h. The system then examines the data from the output stages. Should an error be recognized during any of the monitoring operations described above, the ABS switches off and the ABS indicator lamp comes on.

Diagnosis
ABS can respond to a positive error diagnosis with system switch-off, either immediately or upon termination of the currently active braking control. The error is also stored in the fault memory before switch-off. The access to error data for service workshops is among the features provided by the diagnosis interface (ISO/DIS Standard 9141).

Version ABS 5.3
ABS 5.3 has been designed for vehicles with smaller brake systems; it provides the same range of functions as ABS 5.0, but in a much more compact package. The differences in the respective weights and volumes of the two systems are thus substantial. The reductions in installed length – an important factor – and in weight are achieved by using a much shorter servomotor in combination with a more compact component layout.
The solenoid valves are of separated design, the hydraulic components being located in the hydraulic pressure modulator, and the electrical components (coils) in the plug-on ECU module.
The ECU can be in the form of an installation module for direct mounting on the hydraulic modulator (with plug socket), or it can be a separate unit with a cable connection. The electronic components in the plug-on ECU are of hybrid design. Service replacements can be carried out with minimal effort.

Automatic brake-force differential lock ABS/ABD5
The ABS/ABD5 automatic brake-force differential lock is an extension of ABS 5.0 and ABS 5.3. This system supplements the standard ABS functions by furnishing active braking intervention to improve stability, steerability, and traction during initial acceleration on road surfaces characterized by varying traction on the left and right sides. ABS/ABD5 incorporates the following:
– ABS return concept with closed brake circuits,
– active ABD pressure buildup, achieved by using the self-priming return suction pump to extract brake fluid from the brake master cylinder, and
– redundant computer elements featuring comprehensive monitoring software.

With ABS/ABD5 it must always be possible to control the driven-wheel brakes individually. For this reason, apart from the solenoid valves for generating ABD pressure, the hydraulic modulators in both brake circuits incorporate eight solenoid valves for modulating wheel-brake pressure.
The system responds to incipient spin at one of the driven wheels by applying pressure to the relevant wheel-brake cylinder via the ABS solenoid valves, and triggering a switching valve to isolate the wheel-brake cylinder from the brake master cylinder. The self-priming suction return pump extracts brake fluid from the brake master cylinder through the open suction valve and pumps it to the wheel-brake cylinder where it is required.
The ECU initiates the ABD closed-loop control process whenever a driven wheel starts to display excessive slip during vehicle acceleration:
Braking force is applied to one wheel only – that with the greatest tendency to spin. This is determined by comparing the speeds at the two driven wheels. Once a defined speed differential is exceeded, the brakes are applied at the wheel with the higher rotation rate. The essential speed differential employed for initiating control during initial acceleration – is defined with reference to the vehicle type and drive system (approx. 8...32 km/h). The differential (approx. 6 km/h), is lower at higher vehicle speeds, bringing it into the ideal slip range.

ABS components

Wheel-speed sensors

The ECU uses the signals (wheel fre-quencies) provided by the wheel-speed sensors (Fig. 1) as the basis for deter-mining the wheels' rotational speeds.
The wheel-speed sensor's pole pin with its external winding is located directly above the sensor ring, which is a type of pulse rotor joined directly to the wheel hub (the whcel-speed sensor is also installed in the differential in some appli-cations). The pole pin is connected to a permanent magnet producing an electri-cal field that extends outward to the sen-sor ring. As the ring turns, the pole pin is exposed to an alternating progression of teeth and gaps. This results in the magnetic field changing continuously so that a voltage is generated in the sen-sor's winding, the frequency of which provides a precise index of the current wheel speed.

Various pole-pin configurations (Fig. 2) are available for the different installation conditions in the wheel-hub vicinity. The chisel-type pole pin, designed for vertical installation perpendicular to the sensor ring, is the most common. The rhombus-type pole pin for axial installations is in-stalled at a radial angle to the ring. With both of these designs, it is essential that the pole pin be precisely positioned rela-tive to the sensor ring. On the other hand, no special alignment procedures are re-quired with the round pole pin; however, the sensor ring must have a certain criti-cal diameter or be equipped with fewer teeth. The gap between the wheel-speed sensor and the sensor ring is only about 1 mm; and precise tolerances must be maintained to ensure reliable signal gen-eration. Rigidly mounted wheel-speed sensors also prevent oscillations in the vicinity of the wheel brake from distorting the sensors' signals. Because they are also exposed to water and contami-nation, the wheel-speed sensors are given a coating of grease prior to installa-tion.

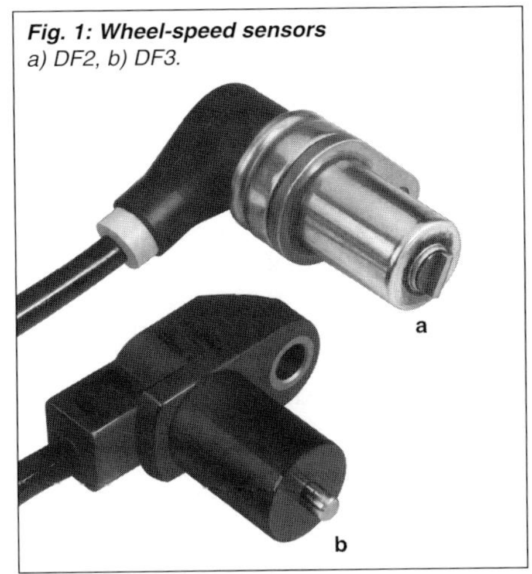

Fig. 1: Wheel-speed sensors
a) DF2, b) DF3.

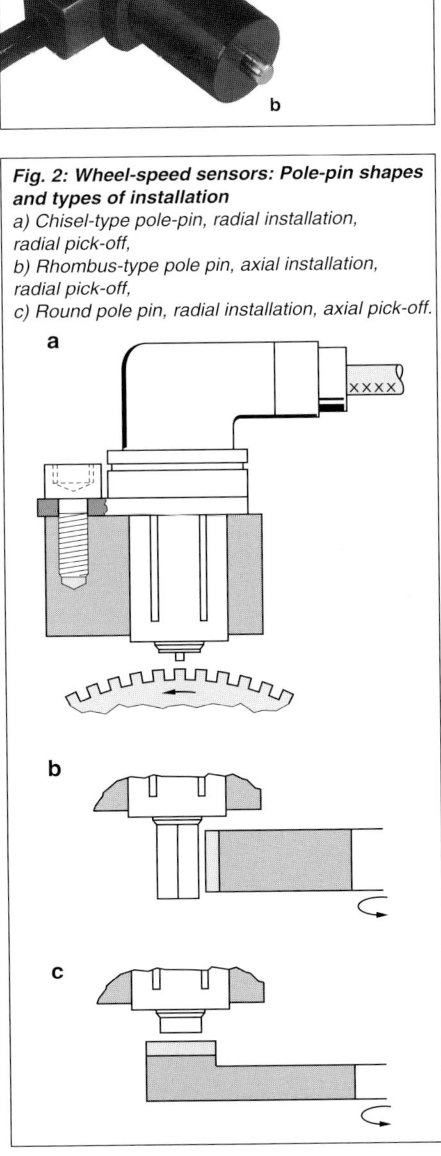

Fig. 2: Wheel-speed sensors: Pole-pin shapes and types of installation
a) Chisel-type pole-pin, radial installation, radial pick-off,
b) Rhombus-type pole pin, axial installation, radial pick-off,
c) Round pole pin, radial installation, axial pick-off.

DF2 wheel-speed sensor

The DF2 wheel-speed sensor (Fig. 3a) consists of individual subassemblies (modules) designed to allow individual testing. This sensor is enclosed within a stainless-steel sleeve, onto which the mounting bracket and its orifice are soldered. The tip of the unit is spray-coated with plastic, and a seal insulates it and its connection from the housing.

DF3 wheel-speed sensor

The DF3 wheel-speed sensor (Fig. 3b) is a simplified version of the DF2, with which it shares both its basic inner structure and its inductive operating principle. The sensor element within the plastic housing is completely encased in resin for protection against external influences. The unit is mounted using a brass sleeve embedded in the housing.

Fig. 3: Wheel-speed sensors (section)
a) DF2 wheel-speed sensor with chisel-type pole-pin,
b) DF3 wheel-speed sensor with round pole pin.
1 Electric cable, 2 Permanent magnet, 3 Housing, 4 Winding, 5 Pole pin, 6 Sensor ring.

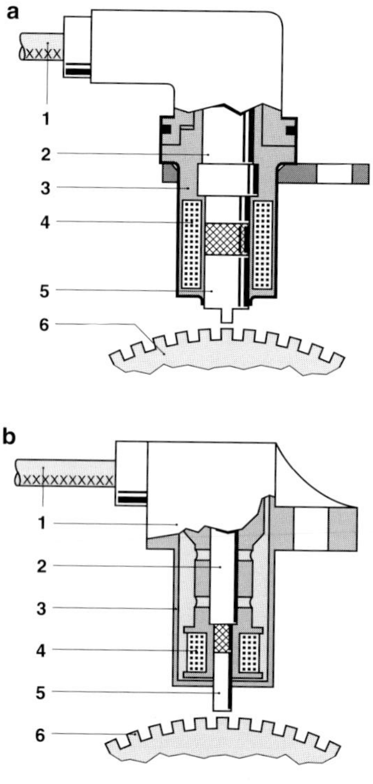

Electronic control unit (ECU)

The electronic control unit receives, amplifies and filters the sensor signals to determine velocities which serve as the basis for calculations of reference speed, brake slip and the wheels' peripheral rates of acceleration and deceleration.

Control unit for ABS 2S

The ECU is an extremely compact package. The individual function modules are highly integrated hybrid devices designed specifically for individual vehicular applications. The computer consists of two digital LSI circuits. These circuits feature discrete semiconductor elements for filtration, signal-strength regulation, reference-pulse generation and interference suppression, as well as incorporating the power transistors used to control the solenoid valves.

The preferred installation location for the ECU is in the passenger compartment, where it is protected from such factors as extreme high temperatures and water spray. Control units intended for installation inside the engine compartment must be specially equipped to withstand more aggressive conditions. Figure 4 is a schematic block diagram illustrating the basic design concept used for the ECU in a 4-channel ABS system:

Input circuit

The input circuit consists of a low-pass filter and an input amplifier. This circuit suppresses interference while at the same time amplifying all signals received from the wheel-speed sensors (channel 1...4). The input circuit also converts the sinusoidal AC voltage from the wheel-speed sensors into square-wave output signals, using these processed signals to control both of the digital controller's LSI circuits.

Digital controller

The digital controller consists of two identical but separate LSI circuits. These circuits operate in parallel, and each processes information from two wheels (Channels 1 and 2, and 3 and 4) and carries out the corresponding logical processes.

This design maintains separate channels to obviate the possibility of central errors in this section of the circuit. It also reduces response delay (dead time) in signal processing to an absolute minimum, with attendant positive consequences for the control process.

The analog square-wave signals representing wheel frequency are converted to 10-bit words in the input sections of the two LSI circuits. This is where interference from factors such as suspension oscillation and road-surface bumps and unevenness are screened out.

Further downstream in the circuit is a serial arithmetic-logic processor. This device uses the processed wheel speed (wheel frequency) as the basis for calculating the controlled variables "slip" and "deceleration" (or "acceleration") for the wheels. A complex controller logic featuring adaptive response for variable functions (i.e., the unit automatically adapts to changing conditions in the controlled system) converts these control signals into positioning commands for the solenoid valves. Communications between the two LSI circuits are maintained through a serial interface; which is connected to the input stage, arithmetic processor and controller logic by means of data links (Fig. 5).

Yet another function module consists of the monitoring circuitry used for error recognition and evaluation. Because ABS exercises direct control over the operation of the service brakes, it is imperative that this circuit maintain extremely high performance levels in the

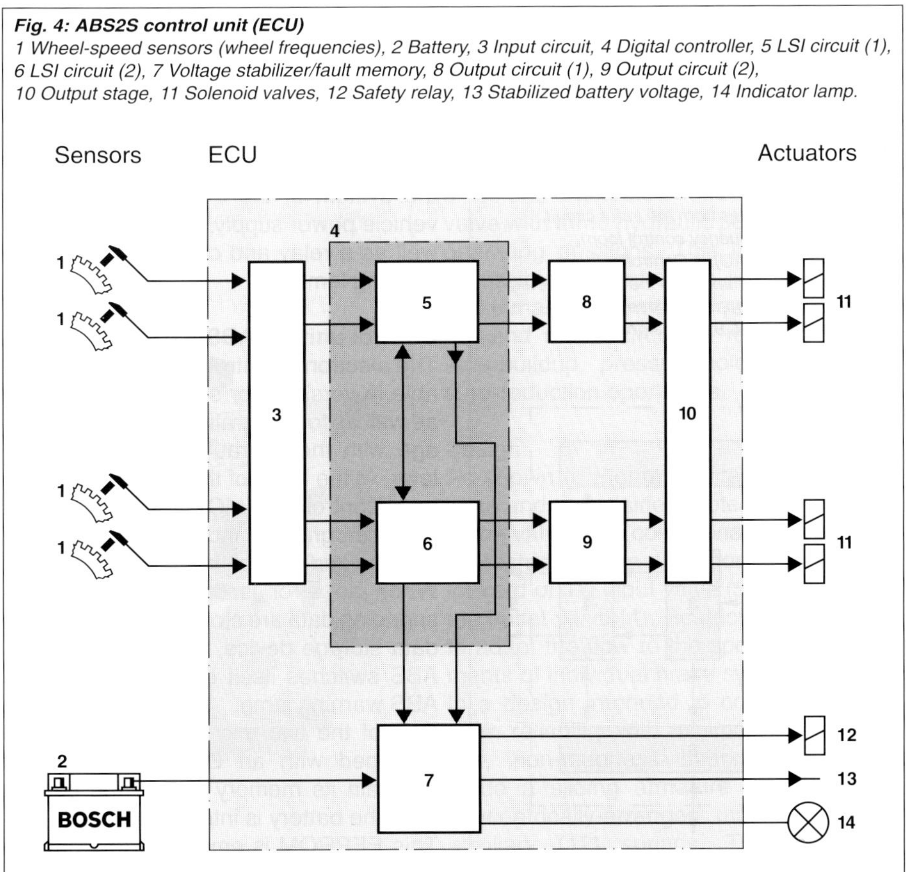

Fig. 4: ABS2S control unit (ECU)
1 Wheel-speed sensors (wheel frequencies), 2 Battery, 3 Input circuit, 4 Digital controller, 5 LSI circuit (1),
6 LSI circuit (2), 7 Voltage stabilizer/fault memory, 8 Output circuit (1), 9 Output circuit (2),
10 Output stage, 11 Solenoid valves, 12 Safety relay, 13 Stabilized battery voltage, 14 Indicator lamp.

open the outlet (4) valves is approximately 0.25 mm; the valve body (9) must therefore be carefully adjusted during assembly. Sealing is provided by steel pellets soldered onto carrier plates (11). These must be extremely small to deal with the high levels of hydraulic force to which they are exposed. The hardened seal seats are machined with extreme precision to ensure efficient sealing at pressures of up to 200 bar.

Constant magnetic response is essential if the required switching times of just a few milliseconds are to be maintained; the components in the solenoid circuit must be manufactured with corresponding precision. A bearing ring (3) with precisely-defined response characteristics is inserted in the valve housing to divert the magnetic field surrounding the coil winding into the armature (6), whence it flows onward to the working air gap (a). The pressure joint must remain sealed at pressures of up to 350 bar (Fig. 7).

Another important factor is consistent application of magnetic forces throughout the armature's stroke. This is achieved with a recess step (14) at the working gap, designed to accommodate the armature's end face. Figure 8 illustrates

the progressions of solenoid forces for holding and for maximum current, as a function of armature travel. A synthetic coating is sprayed onto the winding (7) to protect it against contamination from aggressive brake fluid.

A check valve (8) is installed parallel to the input valve (5). When the brake is released, this check valve opens a supplementary, large-diameter passage leading from the wheel-brake cylinder to the brake master cylinder. This facilitates rapid pressure reductions. It also ensures that it remains possible to release the brake in the event of a (theoretically possible) defect such as a broken main spring or sticking armature.

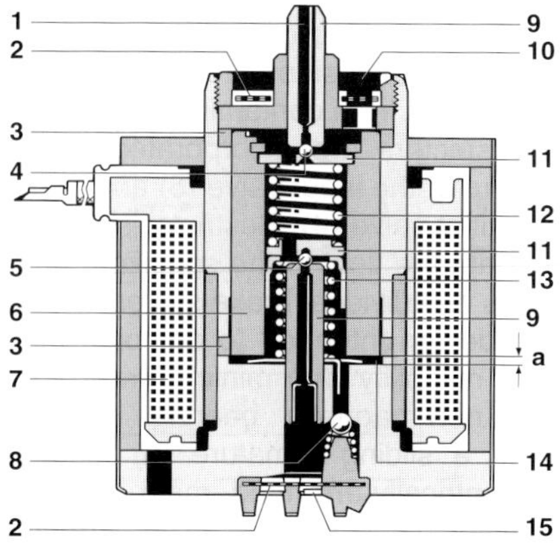

Operation:

1. Pressure buildup phase: When the solenoid is at its base (non-energized) position, the passage between the ports at the brake master cylinder (15) and the wheel-brake cylinder (10) remains unobstructed, allowing unrestricted braking-pressure buildup during normal braking or in the appropriate phase when ABS is active (Fig. 7). In this position the two springs in the solenoid valve – the main (13) and auxiliary (12) springs – exert mutually opposed forces. Because the main spring is under higher tension than the auxiliary unit, the resulting force opens the input valve (5).

2. Pressure-holding phase: The input valve (5) must react to incipient wheel lock by interrupting the connection between the brake master cylinder and the wheel-brake cylinders of the wheels concerned in order to prevent additional increases in pressure. The system applies 50% current (holding current) to winding (7) at the wheel(s). The armature (6) responds by shifting until the input valve (5) is closed by the pellet. The auxiliary spring (12) ceases to exert pressure against the main spring (13) in this position. Because the winding (7) is not able to overcome the pressure exerted by both springs, the armature comes to a halt and remains in this intermediate position.

Fig. 7: ABS2S 3/3 solenoid valve
1 To the return line, 2 Filter, 3 Non-magnetic bearing ring, 4 Outlet valve, 5 Input valve, 6 Armature, 7 Winding, 8 Check valve, 9 Valve body, 10 To the wheel-brake cylinder, 11 Carrier plate, 12 Auxiliary spring, 13 Main spring, 14 Recess step, 15 From brake master cylinder. a) Working air gap.

All three ports are now isolated from one another. A degree of "stroke overlap" (Fig. 8) ensures that the input valve (5) closes before the outlet valve (4) can open.

3. Pressure-reduction phase: In this phase the system must release the excess braking pressure by opening a passage between the affected wheel-brake cylinder and the return line (1) or accumulator.
Maximum current is applied to the winding (7). This allows the armature (6) to overcome the force exerted by the springs (12 and 13) and open the outlet valve (4).
Once sufficient pressure has been discharged from the wheel-brake cylinder, the solenoid valve reverts to either the pressure-holding position or to its (non-energized) basic (pressure-reduction) position, according to momentary requirements.

Hydraulic pressure modulator for ABS 5.0

This is a modular-system hydraulic pressure modulator (Fig. 9). Its basic design enables it to be combined with a plug-on ECU (Fig. 10).

Fig. 9: ABS5.0 (4-channel) modular-system hydraulic pressure modulator
1 Solenoid valve, 2 Pump element,
3 Damping chamber, 4 Accumulator chamber.

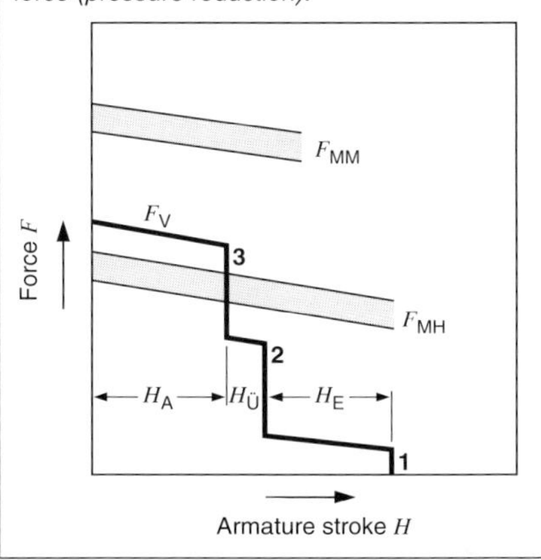

Fig.8: Solenoid-valve forces
F_{MH} *Magnetic force for holding current,*
F_{MM} *Magnetic force for maximum current,*
F_V *Actuating force,* H_A *Discharge stroke,*
H_E *Intake stroke,* $H_Ü$ *Overlap stroke.*
1 Main-spring force minus auxiliary-spring force (pressure buildup), 2 Main-spring force (pressure hold), 3 Main spring force plus auxiliary-spring force (pressure reduction).

Fig. 10: ABS5.0 hydraulic pressure modulator with integrated ECU
1 ECU, 2 Hydraulic pressure modulator.

Return pump

The pump elements are installed in the center of the hydraulic pressure modulator. The electric drive motor for the return pump is located on the side opposite the solenoid valves. This self-priming return pump transfers the brake fluid emerging from the wheel-brake cylinder in the pressure-reduction phase through accumulators and damper chambers on its way back to the brake master cylinder. It serves as the energy source for active braking intervention.

Accumulators and damper chambers

The accumulators and damper chambers are located in the lower section of the hydraulic pressure modulator. The accumulators absorb the surge in brake fluid that accompanies the pressure-reduction process. The damper chambers suppress pressure oscillations within the hydraulic system, preventing them from being propagated back to the brake pedal. They also reduce noise levels.

2/2 solenoid valves

Either three or four pairs (inlet and outlet) of these valves are located in the upper section of the hydraulic modulator – the exact number depending on whether the modulator is installed in a 3 or 4-channel ABS assembly. The operative distinctions are between 3 and 4-channel versions for use in II-pattern brake circuits, and a 4-channel version for X-pattern circuits (Figures 9 and 11).

The solenoid valves are responsible for modulating the pressure in the wheel-brake cylinders during active ABS control.

Hydraulic unit for ABS/ABD5

Hydraulic modulators for ABS/ABD5 systems and II-variant braking-force distribution include the following additional components:
– a pilot valve (USV) for reverting from braking to ABD operation, featuring an integral pressure-relief valve (DBV) for limiting the ABD system pressure,
– a suction valve (ASV) to open a passage between the self-priming return pump and the brake master cylinder port during ABD operation, and
– a modified return pump with a self-priming circuit.

For X-variant braking-force distribution, the system is expanded to include the following:
– two pilot valves (USV) for reverting from braking to ABD operation, featuring two integral pressure-relief valves (DBV) for limiting ABD system pressure,
– two suction valves (ASV) to open a passage between the self-priming return pumps and the brake master cylinder ports during ABD operation, and
– a return pump (modified from the ABS hydraulic pressure modulator) with two self-priming circuits.

Fig. 11: ABS5 2/2 solenoid valves
a) Intake valve.
De-energized = open,
b) Discharge valve.
De-energized = closed.

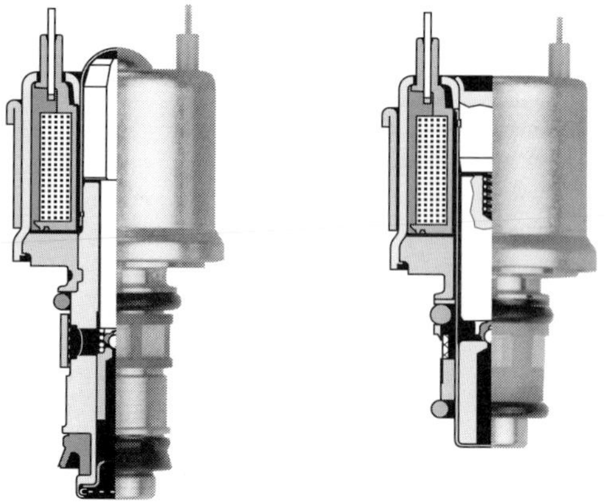

Electrical circuits

Figure 12 shows a circuit diagram with the electrical circuit for a 4-channel ABS 2S unit equipped with four wheel-speed sensors and four solenoid valves. The electrical connections linking the various electrically-controlled components are gathered in a wiring harness. Both the wiring harness and the components themselves must be installed in protected locations to prevent water damage. Another hazard to avoid is corrosion, with its resulting higher levels of contact resistance and attendant operating difficulties, which ultimately lead to system failure.

To ensure that the system operates at maximum efficiency, the design process for each component includes wiring calculations to evaluate the line losses in critical connections. Excessive voltage drop (relative to the ECU) could lead to premature system switch-off. Line losses can also affect operation of the solenoid valves and return pump, with switching delays and reduced pumping efficiency as the results.

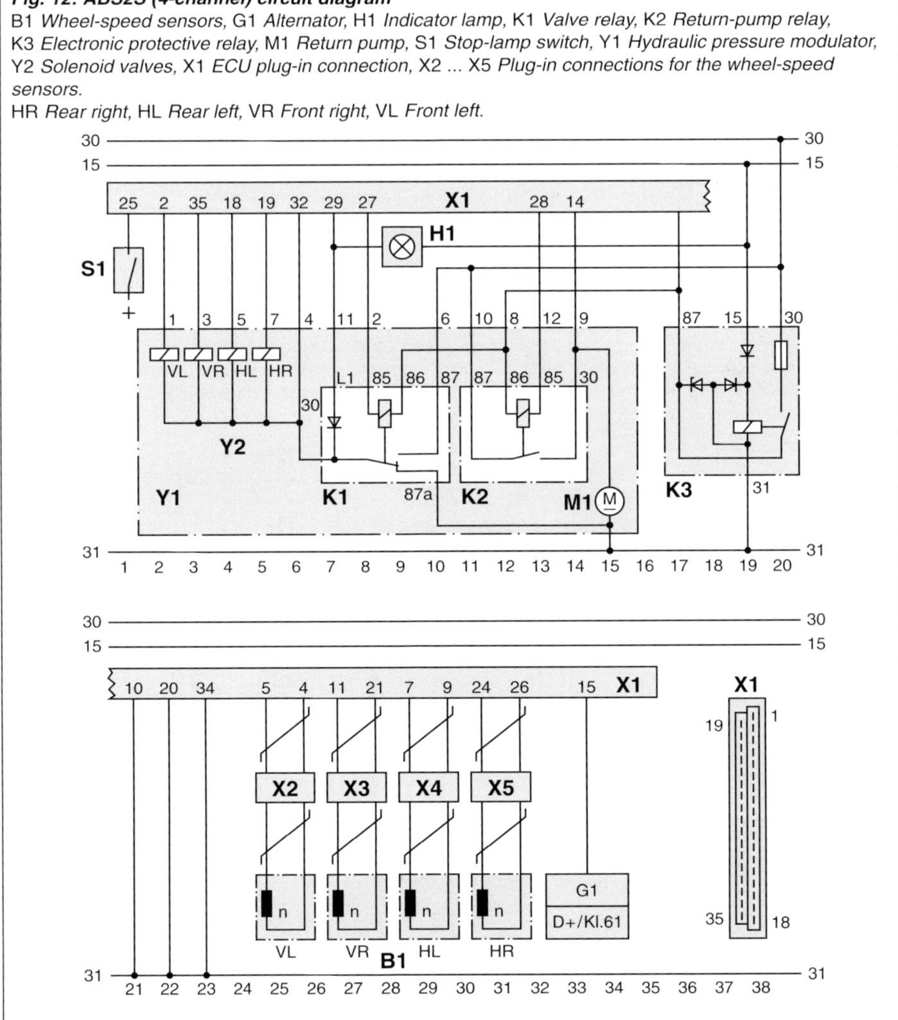

Fig. 12: ABS2S (4-channel) circuit diagram
B1 *Wheel-speed sensors,* G1 *Alternator,* H1 *Indicator lamp,* K1 *Valve relay,* K2 *Return-pump relay,* K3 *Electronic protective relay,* M1 *Return pump,* S1 *Stop-lamp switch,* Y1 *Hydraulic pressure modulator,* Y2 *Solenoid valves,* X1 *ECU plug-in connection,* X2 ... X5 *Plug-in connections for the wheel-speed sensors.*
HR *Rear right,* HL *Rear left,* VR *Front right,* VL *Front left.*

ASR traction control

Critical driving situations are not restricted to braking; they can also occur during standing-start and moving acceleration (especially on slippery gradients) and during cornering. These conditions can present drivers with more than they can handle. The result: Dangerous driving errors.

ASR traction control is designed to solve these problems. The primary purpose of ASR, an expanded version of ABS, is to reduce the demands placed on the driver by maintaining vehicle stability and steering response under acceleration (provided that the physical limits are not exceeded). ASR does this by adapting the engine torque to levels corresponding to the traction available at the road surface – before the situation becomes critical (Fig. 1). By combining ASR and ABS it is possible to obtain higher levels of safety through dual-purpose application of system components.

ASR system requirements

Base system

The ASR traction-control system must be capable of inhibiting wheelspin during initial or moving acceleration under the following conditions:
– when the lane is slippery on one or both sides,
– as the vehicle emerges from iced-over parking lots and highway shoulders,
– during acceleration when cornering, and
– when starting off on a gradient (closed-loop control of applied tractive force based on regulation of brake pressure at the potentially spinning wheel).

The traction-control system must also intervene in the following situations:
– When a wheel spins – the same as when it locks – the lateral forces which it can transmit are limited; the vehicle becomes instable and the rear fishtails. ASR maintains vehicle stability for enhanced safety.

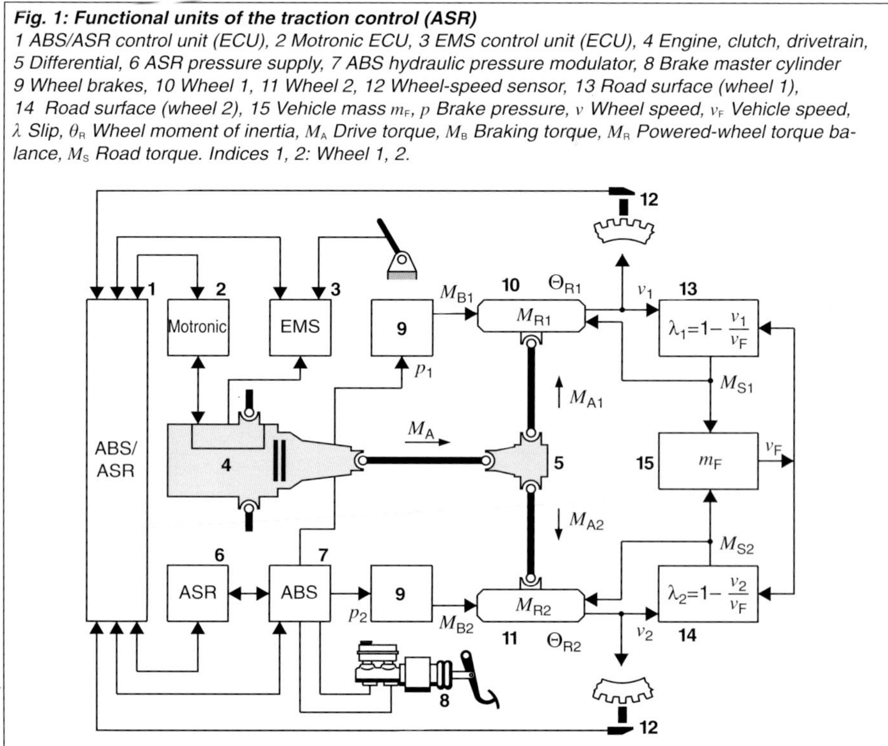

Fig. 1: Functional units of the traction control (ASR)
*1 ABS/ASR control unit (ECU), 2 Motronic ECU, 3 EMS control unit (ECU), 4 Engine, clutch, drivetrain,
5 Differential, 6 ASR pressure supply, 7 ABS hydraulic pressure modulator, 8 Brake master cylinder
9 Wheel brakes, 10 Wheel 1, 11 Wheel 2, 12 Wheel-speed sensor, 13 Road surface (wheel 1),
14 Road surface (wheel 2), 15 Vehicle mass m_F, p Brake pressure, v Wheel speed, v_F Vehicle speed,
λ Slip, θ_R Wheel moment of inertia, M_A Drive torque, M_B Braking torque, M_R Powered-wheel torque balance, M_S Road torque. Indices 1, 2: Wheel 1, 2.*

– Spin-slip also leads to increased tread wear and drivetrain stress (e.g., differential). ASR avoids the drivetrain loads that occur when a spinning wheel suddenly finds traction on a high-adhesion surface.

– ASR must be ready to intervene automatically at all times. ASR employs the difference in slip rates at the drive wheels to distinguish between cornering and acceleration slippage. The tires do not "drag" in tight-radius corners, as can occur with a differential lock. Limited-slip and locking differentials cannot always inhibit wheelspin resulting from excess throttle applications. In contrast, ASR also regulates the engine output to ensure that the wheels retain traction.

– The system responds to conditions in the physical threshold range by triggering a warning lamp to alert the driver.

MSR engine drag-torque control

ASR can be expanded and complemented with the addition of a supplementary MSR engine drag-torque control device. When shifting down, or when the throttle closes suddenly on low-traction road surfaces, the engine's braking effect can lead to excessive braking slip at the drive wheels.

MSR responds by initiating a small increase in the engine's torque to reduce the braking effect at the wheels to a level commensurate with vehicle stability.

EMS electronic engine-power control

ASR must be able to intervene irrespective of the driver's throttle input. It is therefore necessary to replace the mechanical linkage between accelerator pedal and throttle valve (on gasoline engines), or between the pedal and the control lever on the injection pump (on diesel engines), with EMS "electronic engine-power control" (alternatively known as "electronic gas pedal," "E-Gas," "drive-by-wire," etc.). EMS assigns ASR control commands priority over driver inputs.

A pedal-travel sensor converts the position of the accelerator pedal into an electrical signal. The EMS control unit

Fig.2: Electronic engine-power control (EMS) for ASR
1 ABS/ASR control unit (ECU), 2 EMS control unit (ECU), 3 Accelerator pedal, 4 Servomotor, 5 Throttle valve or 6 Diesel fuel-injection pump.

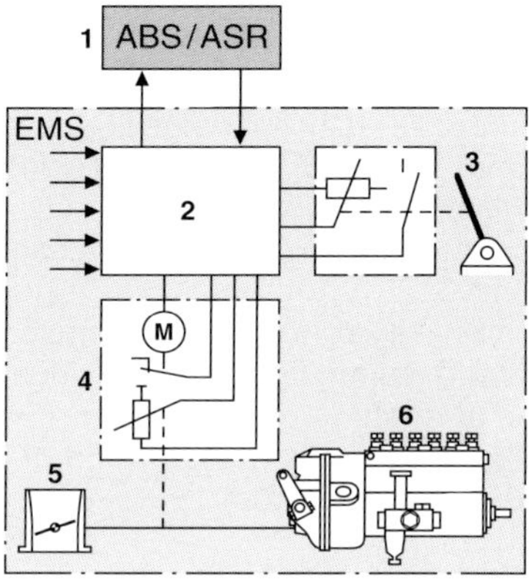

consults programmed factors and signals from other sensors (e.g., temperature, engine speed) in converting this pedal signal into a control voltage for the electric servomotor. The servomotor actuates the throttle valve (or pump control lever on diesels) and relays the position back to the ECU (Fig. 2).

Operation

When the driver depresses the accelerator, engine torque and the resulting drive torque both increase. If the road surface is capable of providing adequate "support" for this increased torque, then the vehicle can accelerate without restriction.

However, at least one of the drive wheels will start to spin as soon as the drive torque exceeds the physical maximum that can be applied through the road. The result is a reduction in effective tractive force, while the attendant loss of lateral adhesion leads to vehicle instability. Within fractions of a second, ASR responds by regulating drive-wheel slippage down to the optimum level.

As Figure 1 illustrates, variations in torque balance M_R can be employed to influence the monitored wheel speeds v_1

and v_2 – and with them the drive slip λ – at each of the drive wheels.

The torque balance M_R consists of the drive torque M_A, braking torque M_B and the road-surface factor M_S (representing the torque that can be transferred to the road surface).

On vehicles with spark-ignition engine, the drive torque M_A is regulated using the following systems:
– throttle valve (throttle-valve adjustment),
– ignition system (advance-angle adjustment, suppression of individual ignition pulses),
– injection system (suppression of individual injection pulses).

On diesel-equipped vehicles the drive torque M_A is regulated by adjusting the control lever on the injection pump (reduction of injected fuel quantity).

The brake system can be used to control the braking torque M_B at the individual wheels. For this, the ABS hydraulic system must be expanded accordingly.

Figure 3 compares the response delays associated with various types of ASR intervention in the engine-management. As the illustration shows, in the case of 2-wheel-drive vehicles, the relatively extended response delays mean that it is not possible to obtain satisfactory results with control based exclusively on throttle-valve adjustments.

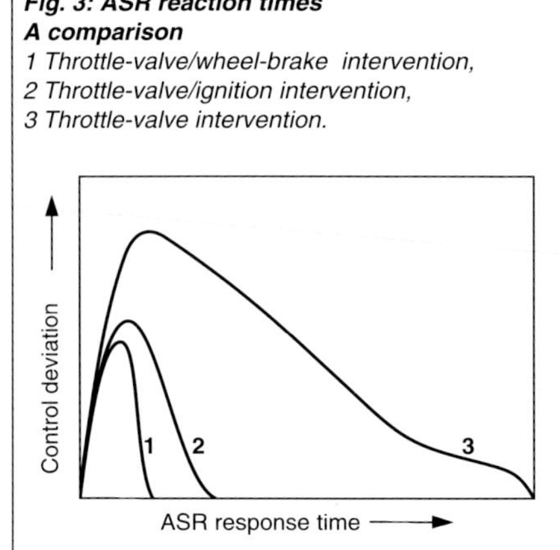

Fig. 3: ASR reaction times
A comparison
1 Throttle-valve/wheel-brake intervention,
2 Throttle-valve/ignition intervention,
3 Throttle-valve intervention.

ASR versions

The following is limited to descriptions of systems for vehicles with spark-ignition engine.

ASR2-DKB with engine and braking intervention

This version is composed of ABS2S componentry along with ASR featuring combined throttle-valve (Drosselklappe, or DK) control and braking (B) intervention. Each of the ABS electronic and hydraulic systems is expanded to include the ASR components (Figures 4 and 5).

Engine intervention

Engine intervention with throttle-valve control is provided by the EMS "electronic engine-power control."

Braking intervention

An effective and convenient means for implementing ASR control of braking forces is available in the form of the components already installed for ABS. A supplementary pilot valve (11) switches from normal brake-system operation to ASR control for the duration of the operation. This makes it possible for the fully-charged brake-fluid accumulators (15) to apply braking pressure to the wheel-brake cylinders at the powered wheels (1) without any corresponding movement at the brake pedal. As with ABS, pressure-modulation is carried out by the solenoid valves (5...8) with their three positions for pressure buildup, pressure holding and pressure release. Pulse control of the solenoid valves is provided by the ABS/ASR control unit whenever the control system recognizes incipient wheelspin at the powered wheels. The return pump (3) continues to operate for the entire duration of the traction-control process in order to help discharge pressure from the drive wheels' wheel-brake cylinders. The discharged brake fluid flows back to the accumulator (15). A charging pump (14) refills the accumulator when ABS/ASR is inactive.

The braking intervention also provides a form of slip-limitation or locking function at the differential. It thus combines improved vehicle stability and steering control with more effective application of accelerative forces, especially on surfaces affording different traction levels on left and right sides. Figure 6 illustrates the linear forces applied to the drive wheels. The wheel on the surface with the higher coefficient of friction μ_h is able to transfer the higher force F_h ("h" for high), the wheel on the area with the lower friction coefficient μ_l transmits the lower force F_l ("l" for low). At the same time, the differential will only permit a total force transfer of $2 \cdot F_l$ on this type of heterogeneous surface. Thus brake force F_B must be applied to prevent the wheel on the μ_l surface from spinning in response to excessive drive torque. The vehicle's maximum tractive force can then be effectively applied:

$$F_{total} = F_h + F_l = 2 \cdot F_l + F_B{}^*$$

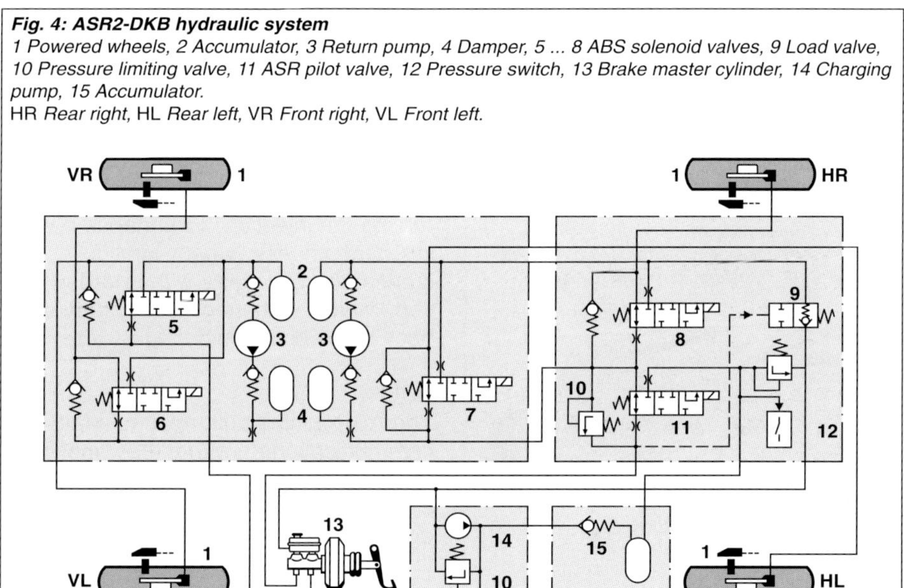

Fig. 4: ASR2-DKB hydraulic system
1 Powered wheels, 2 Accumulator, 3 Return pump, 4 Damper, 5 ... 8 ABS solenoid valves, 9 Load valve, 10 Pressure limiting valve, 11 ASR pilot valve, 12 Pressure switch, 13 Brake master cylinder, 14 Charging pump, 15 Accumulator.
HR *Rear right,* HL *Rear left,* VR *Front right,* VL *Front left.*

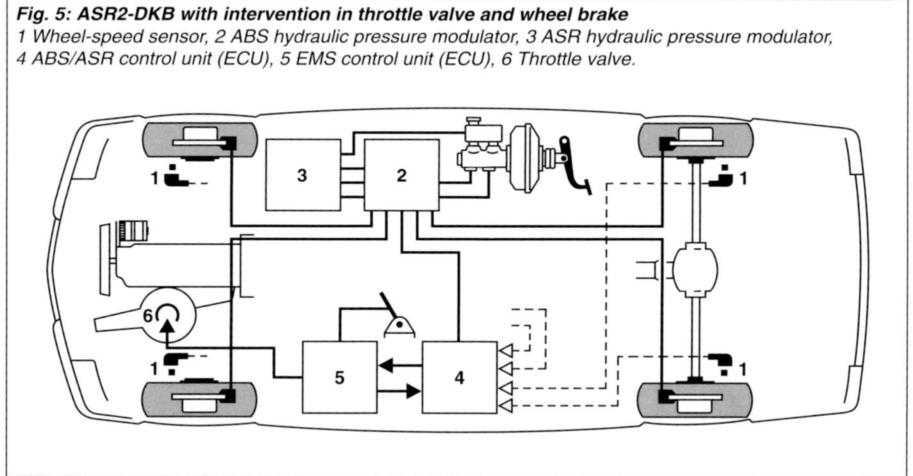

Fig. 5: ASR2-DKB with intervention in throttle valve and wheel brake
1 Wheel-speed sensor, 2 ABS hydraulic pressure modulator, 3 ASR hydraulic pressure modulator, 4 ABS/ASR control unit (ECU), 5 EMS control unit (ECU), 6 Throttle valve.

$F_B{}^*$ is derived from F_B and takes into account the differences in effective radii. The tractive torque is regulated to provide maximum potential drive force. In the process, brief supplementary braking forces may also be applied to the wheel on the μ_h surface (Fig. 6).

ASR2-DKZ/MSR with engine intervention

This version consists of the ABS 2S componentry, together with ASR in which the EMS throttle-valve control function (DK) is supplemented by additional ignition (Z), and injection-system control procedures to reduce the delay before engine-torque reductions become effective.

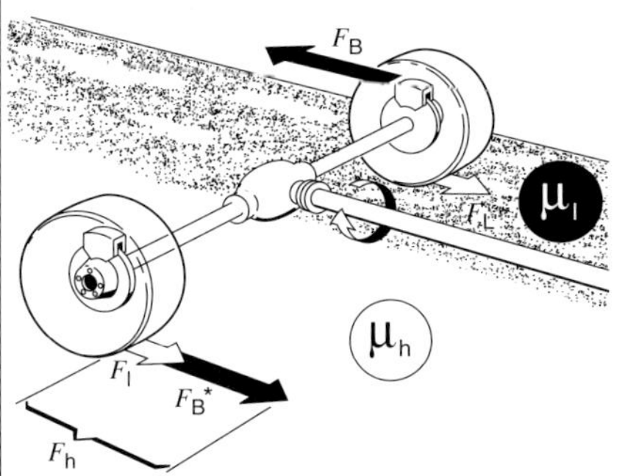

Fig. 6: Limited-slip differential effect resulting from brake intervention
F_B Braking force, $F_B{}^*$ Braking force ref. to radii of action, μ_l Low braking-force coefficient, μ_h High braking-force coefficient, F_l Transferable driving force to μ_l, F_h Transferable driving force to μ_h.

MSR engine drag-torque control is integrated in this system in which the control functions are limited to engine output (no braking intervention). The components in the ABS hydraulic system are unmodified, while the ABS electronic system is expanded to include the ASR circuits (Fig. 7).

Engine intervention

Active intervention in the engine's injection and ignition systems is primarily intended to ensure that the vehicle remains stable. When necessary, the system's initial response is to transmit a "retard ignition" signal through the interface. If, notwithstanding these measures, the drive slip continues to increase or fails to decrease, the unit suppresses the ignition pulses while simultaneously interrupting the injection. To enhance operating smoothness when ignition is resumed, the unit starts with retarded ignition, which it then gradually advances back to the optimum timing.

ECU

The ASR control unit must maintain data communications with the individual ECU's for the electronic engine-power control and with Motronic engine management. A comprehensive interface is thus essential (Fig. 9). A safety circuit monitors the wiring connecting the various control units.

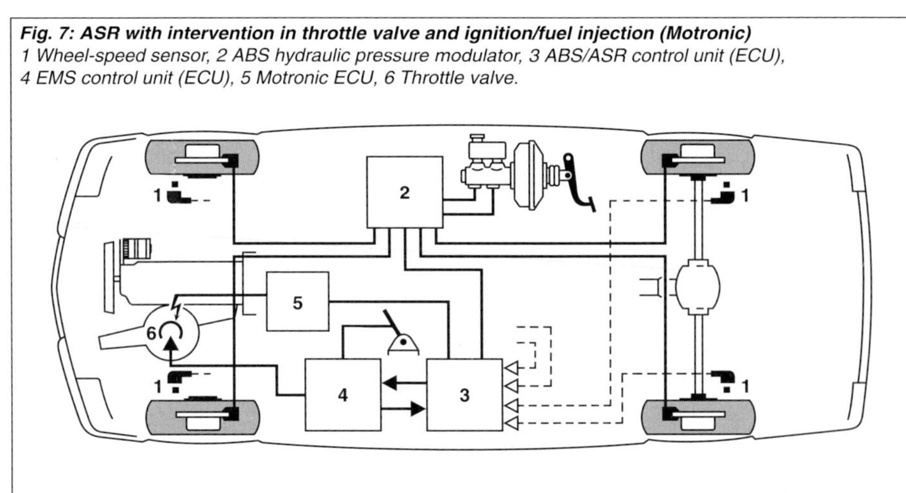

Fig. 7: ASR with intervention in throttle valve and ignition/fuel injection (Motronic)
1 Wheel-speed sensor, 2 ABS hydraulic pressure modulator, 3 ABS/ASR control unit (ECU), 4 EMS control unit (ECU), 5 Motronic ECU, 6 Throttle valve.

ASR5 with engine and braking intervention

ASR5, which is based on the ABS/ABD5 system, supplements ideal levels of vehicle stability and steering response with optimum traction during acceleration from rest.

The intake and output valves in the combined ABS/ASR5 hydraulic unit incorporate additional suction and pilot-valve units (Figures 8 and 10).

In their non-energized states, the intake valve is open (throughflow mode), while the outlet valve is closed. During ASR operation, the suction valve (closed in the non-energized state) opens; this valve features a large-diameter orifice to facilitate extraction of the brake fluid from the brake master cylinder. The pilot valve remains open when no current is applied; for active ASR operation it shifts to its closed position prior to assuming the role of a pressure-relief valve.

Braking intervention

To improve traction – especially on road surfaces with different friction coefficients at crown and shoulder – the system brakes the drive wheel (which shows the greater tendency to slip) to a pre-defined slip rate.

Engine intervention

Various methods are available for reducing the engine's drive torque; the specific requirements of the ABS/ASR system and the type of engine-management system used will determine which are employed:
– adjustment of throttle-valve angle,
– suppression of individual injection and ignition pulses, and
– reduction of ignition advance angle.

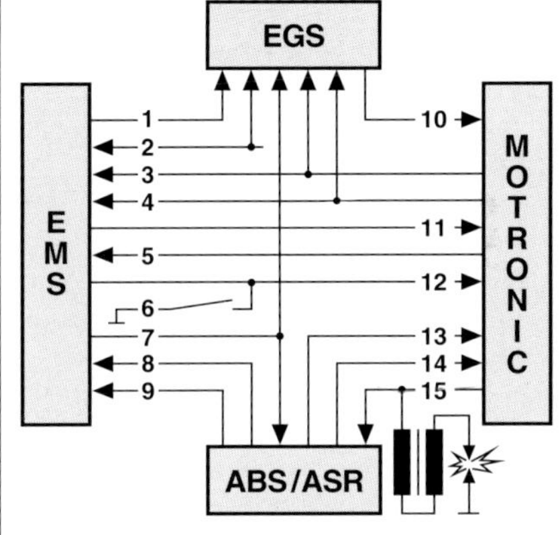

Fig. 9: Interface overview
Control units: EGS (electro-hydraulic transmission-shift control), EMS (electronic throttle control), ABS/ASR Motronic.
1 Kick-down, 2 Drive/program settings, 3 Ignition point, 4 Injection duration, 5 Engine temperature, 6 Stop-lamp switch/accelerator-pedal switch, 7 Throttle-valve input, 8/9 Throttle-valve reduction/increase, 10 Engine intervention, 11 Full-load contact, 12 Idle contact (overrun fuel cutoff), 13 Idle-speed increase (overrun fuel cutoff inactive), 14 Ignition-pulse suppression, 15 Ignition, primary current.

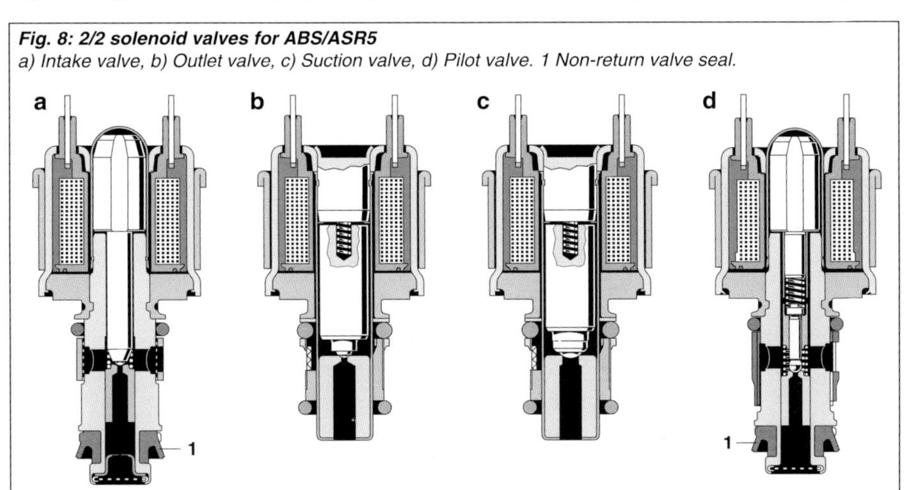

Fig. 8: 2/2 solenoid valves for ABS/ASR5
a) Intake valve, b) Outlet valve, c) Suction valve, d) Pilot valve. 1 Non-return valve seal.

Adjustment of throttle-valve angle

The system employs closed-loop control to adjust the SI-engine's drive torque for optimal slip rates. At low vehicle speeds the drive wheel with the higher friction of coefficient (select-high control) provides the reference; at high vehicle speeds the drive wheel with the lower friction coefient (select-low control).

The system supplements adjustments to the throttle-valve angle – which provide only relatively slow response, especially on vehicles with rear-wheel drive – by rapidly adjusting the ignition angle and suppressing individual ignition and injection pulses.

The base version employs an ASR throttle-valve actuator (ADS with electric servomotor) to reduce the SI-engine's drive torque by decreasing the throttle-valve aperture.

A potentiometer monitors the current throttle-valve angle, and the ECU uses this throttle-valve signal as the basis for calculating subsequent positioning commands for the throttle-valve actuator.

ASR throttle-valve actuator ADS1:

The ADS1 ASR throttle-valve actuator is "spliced" into the Bowden cable between the accelerator pedal and the throttle valve (Fig. 11).

Each segment of the divided Bowden cable terminates in a cable pulley which, in turn, are linked with each other by a tensioned coupling spring. When the driver depresses the accelerator pedal during normal operation, the throttle-valve is opened via coupling spring and cable pulleys.

Upon receiving ASR commands to reduce the throttle-valve aperture, the ECU triggers a servomotor. This uses a gear-set to rotate the cable pulleys which are normally in a state of opposed tension. This results in a change to the Bowden cable's effective length, and therefore in a reduction in the throttle-valve aperture despite the pressure that the driver applies at the accelerator pedal.

As soon as the vehicle regains an adequate level of traction and stability, the ECU transmits a deactivation signal to the servomotor, which responds by returning to its initial position.

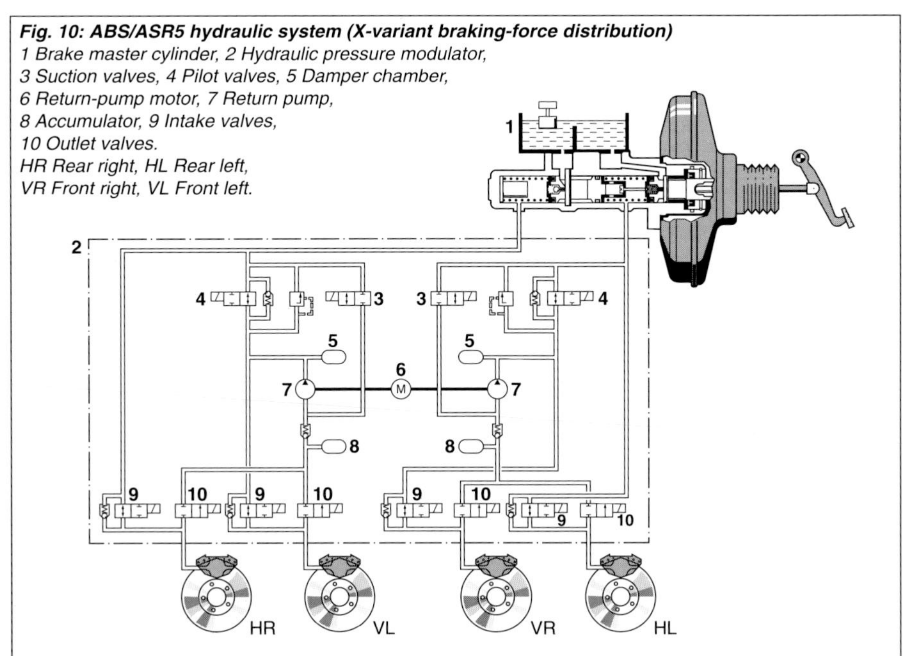

Fig. 10: ABS/ASR5 hydraulic system (X-variant braking-force distribution)
1 Brake master cylinder, 2 Hydraulic pressure modulator,
3 Suction valves, 4 Pilot valves, 5 Damper chamber,
6 Return-pump motor, 7 Return pump,
8 Accumulator, 9 Intake valves,
10 Outlet valves.
HR Rear right, HL Rear left,
VR Front right, VL Front left.

ASR throttle-valve actuator ADS2:
Applications with the ASR throttle-valve actuator ADS2, employ a separate Bowden cable to connect the accelerator pedal with the throttle-valve.
Because the forces opposing movement are smaller than with the ADS1 version, a more compact design can be used. For the same reason, a low-force coupling spring is adequate. The spring is installed directly at the throttle valve, in parallel with the return spring. The feedback which the driver feels at the accelerator pedal with the ADS1, is not present when the ADS2 throttle-valve actuator is used. However, the throttle valve must be modified for operation with the disconnection mechanism.

Ignition and injection suppression:
The system intervenes in the ignition process via an interface to the engine-management system. Intervention options include suppression of individual ignition and/or injection pulses, reduction in the ignition's advance angle and increases in the engine's idle speed.
Ignition-pulse suppression is used to achieve rapid reductions in drive slip, particularly on vehicles with rear-wheel drive. There is only a minimal delay in initiating suppression, and with it the reduction in engine drive torque. The injection pulses are suppressed at the same time to prevent unburned mixture from being discharged into the exhaust system when the ignition pulses are cancelled for a longer period of time.
There is a slight increase in response times when only the injection pulses are suppressed.
Reductions in the ignition advance angle also provide rapid (but smoother) reductions in the engine's drive torque. Adjustments to ignition timing are thus preferable to individual ignition-pulse suppression in all cases in which they can provide adequate results.
Because active intervention is generally limited in duration, any potential consequences for exhaust-gas composition and possible loads on the catalytic converter are not critical factors.
The high-idle feature is intended to respond to incipient loss of drive-wheel stability stemming from engine braking (drag torque); its chief field of application is in rear-wheel drive vehicles with

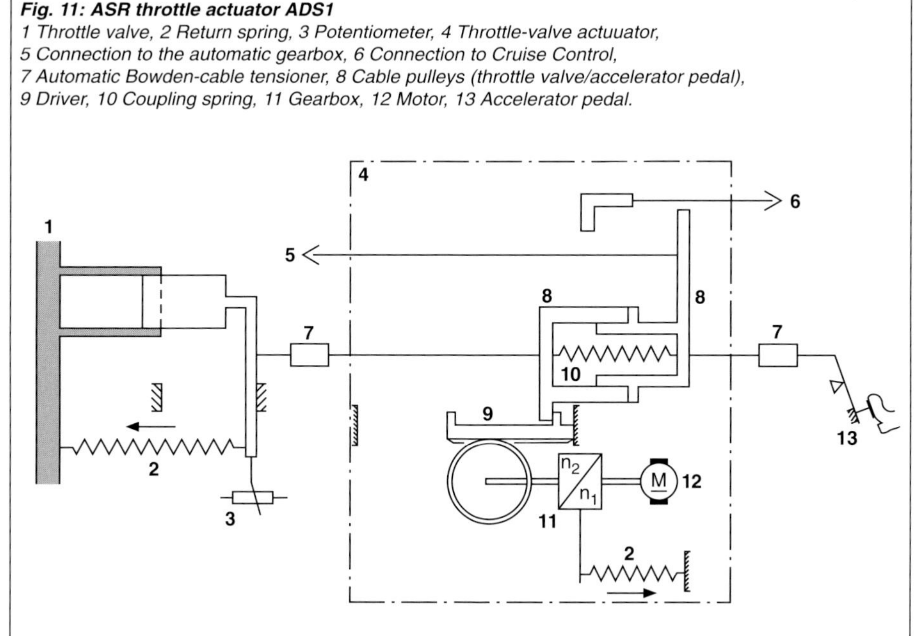

Fig. 11: ASR throttle actuator ADS1
1 Throttle valve, 2 Return spring, 3 Potentiometer, 4 Throttle-valve actuuator,
5 Connection to the automatic gearbox, 6 Connection to Cruise Control,
7 Automatic Bowden-cable tensioner, 8 Cable pulleys (throttle valve/accelerator pedal),
9 Driver, 10 Coupling spring, 11 Gearbox, 12 Motor, 13 Accelerator pedal.

manual transmission. The ASR control unit initiates the process by transmitting a signal to the engine-management ECU for implementation.

Injection suppression with
SEFI-Motronic:
Substantial reductions in overall ASR system complexity can be obtained with a concept based on suppressing individual injection pulses with the SEFI Motronic (Motronic with sequential fuel injection).

Under ideal conditions all of the required components will already be present in the vehicle:
– the ABS/ASR5 control unit,
– an engine-management system with sequential fuel injection, and
– an interface between the two control units.

The wheel-speed sensors, already present as standard ABS components, furnish the basic ASR data in the form of the signals used to determine the speeds of the driven and non-driven wheels.
Bidirectional data communications are maintained between the control units for ABS/ASR and fuel injection: The ABS/ASR control unit responds to excessive wheel slip by transmitting a command for reduced engine drive torque to the fuel-injection ECU.
The data flow in the other direction provides the ABS/ASR control unit with information on driver input and on the current operating status of the engine (e.g., engine speed).
The system achieves precisely graduated control of engine torque by switching the injection at individual cylinders on and off in cycles of two crankshaft rotations, according to the degree of wheel slip.
The result is a mean torque situated between two cylinder levels. The system is thus capable of adjusting the number of active cylinders in increments of "0.5" cylinders.
The system can also use short-duration ignition-pulse suppression to supplement the injection-pulse suppression when rapid reductions at a high initial output level are required.

Control unit (ECU)
The ABS/ASR control unit's input amplifier receives the signals from the wheel-speed sensors and converts them to square-wave signals; the frequency of which serves as an index of wheel speed. The two microprocessors evaluate this data to determine wheel speed and acceleration rates, from which the basic parameters are derived for calculating the specified and actual values used for traction control. All circuits – signal processing, control algorithms (arithmetic operations), and monitoring software – are present in each of the two micro-controllers (system redundancy).
Depending on system version, the ECU for engine and braking intervention may also incorporate supplementary interfaces for such functions as transmitting pulse-width-modulated signals, CAN and control software, as well as output amplifiers for controlling the ASR throttle-valve actuator.

ASR5 with engine intervention
This ABS/ASR5 system (with engine intervention) also provides substantial benefits in the areas of stability (on vehicles with rear-wheel drive) and steerability (on FWD vehicles).

Engine intervention
The ABS/ASR5 system with engine intervention operates by adjusting the engine's drive torque alone; it does not use the kind of braking intervention for supplementary torque adjustment that would be necessary for optimal traction control. This system without braking control requires only the ABS5 hydraulic modulator.

ECU
The ABS/ASR control unit for engine (without braking) intervention dispenses with the electronic components required to trigger brake intervention. The output-amplifier stages used to operate the

traction-control solenoid valves (pilot and suction valves) are not required in this version. In this particular ECU, the ABS control unit includes a supplementary interface for triggering engine-torque intervention. It also monitors a signal from the engine-management ECU, using this data as a basis for determining engine speed for traction control. Here as well, an indicator lamp informs the driver whenever the traction-control system goes into active operation.

Fig.12: Winter-testing the ABS/ASR in North Sweden with different coefficients of friction between the left and right wheels (μ -split).

Commercial vehicles – Basic concepts, systems and diagrams

Basic components

Every brake system can be divided into:
– energy supply,
– actuating device,
– transmission (force-transfer) device, and
– brake.

The actuating device regulates the flow of energy as it travels from the energy-supply device, through the transmission (force-transfer) device, and to the wheel brake.

Figures 1 and 2 show the individual components within these four basic building blocks as found in a compressed-air brake system. Figure 1 provides a schematic diagram of the individual devices, while Figure 2 shows them in the form of symbols.

Energy supply

The energy source supplies the energy required for braking. The options include pneumatic, hydraulic and mechanical energy, as well as the muscular force furnished by the vehicle operator.

The energy-supply system also includes mechanisms designed to control, condition and (in some cases) to store the energy – provided that they are not included in the transmission device.

The brake systems in some commercial vehicles rely on compressed air exclusively. Others use a variety of energy sources. One example would be a truck equipped with compressed-air service and secondary brakes, on which the parking brake is operated using muscular force.

Actuating device

The actuating device comprises those components responsible for actuating and regulating operation of the brake system.

Compressed-air brake systems may feature an actuating device actuated by the vehicle operator (referred to as the "driver" in the following) at a brake pedal or lever. Other types of compressed-air brake system are triggered automatically, e.g., when the connection between tractor and trailer is interrupted.

Transmission (force-transfer) device

The transmission (force-transfer) device includes those system components responsible for relaying the energy to the brake. This system extends from the brake pedal (or lever) to the wheel brake itself. The energy-storage devices for the individual brake circuits (including safety devices) are also classified as part of the transmission device.

Brakes

The brake is the part of the brake system in which the forces that oppose and counteract vehicle movement are generated. The brake operates by converting the vehicle's kinetic energy into heat.

Friction brakes

Most vehicles are equipped with friction brakes. Friction brakes use the braking energy from the energy-storage device to generate tension; this tension presses the friction linings (or pads) against the brake drums (or discs).

During braking, friction is generated by pressing the non-rotating brake linings/

Fig. 1: Basic components of a brake system. Individual devices shown as drawings.
A Energy supply.
1 Air compressor, 2 Pressure regulator, 2a Air drier, 2b Regeneration-air tank.
B Actuating device.
3 Service-brake valve 4 Parking-brake valve.
C Transmission device.
5 4-circuit protection valve, 6 Air reservoir, 7 Drain valve, 8 Relay valve, 9 Load-sensing valve,
10 Brake cylinder, 11 Combination brake cylinder, 12 To brake.

VA *Front axle*, HA *Rear axle*.

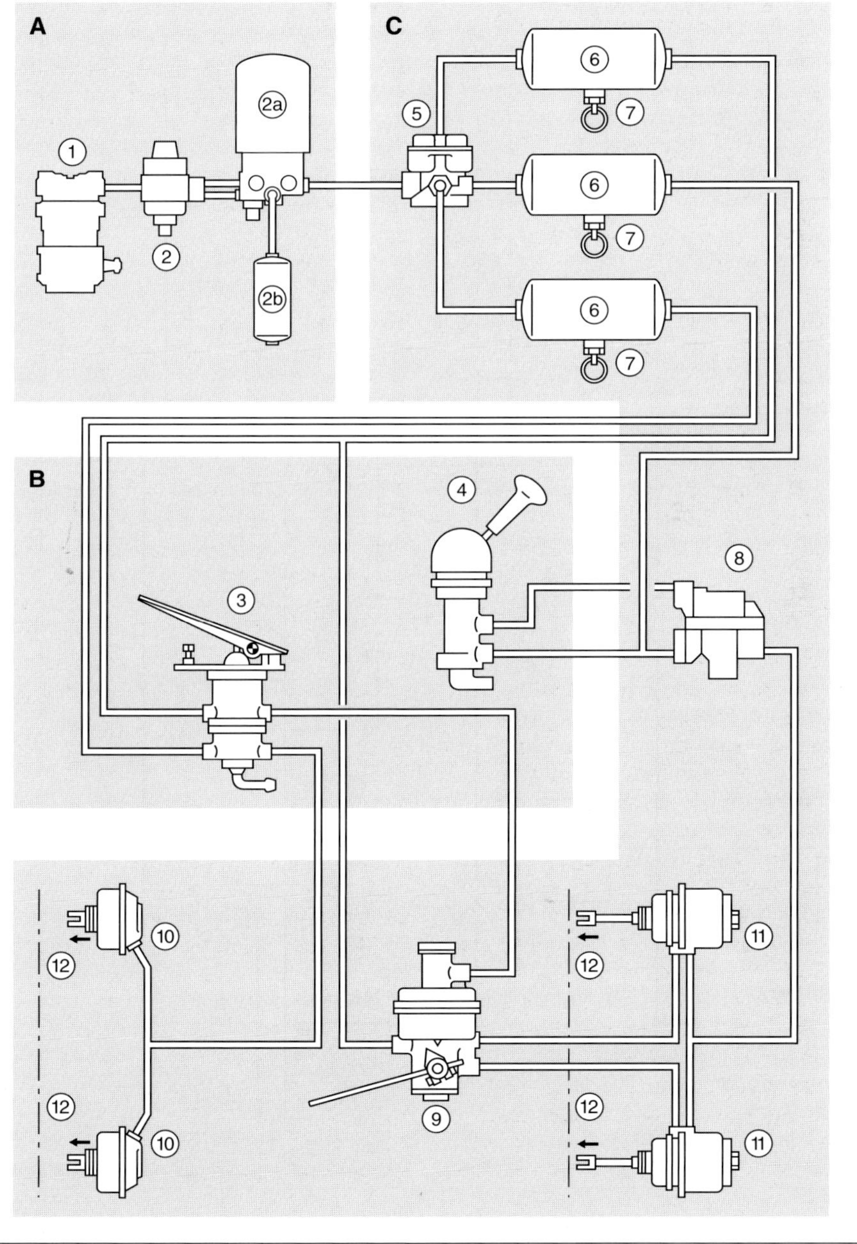

Fig. 2: Basic components of a brake system. Individual devices shown as drawings.
A Energy supply.
1 Air compressor, 2 Air drier with pressure regulator and regeneration-air tank.
B Actuating device.
3 Service-brake valve, 4 Parking-brake valve.
C Transmission device.
*5 4-circuit protection valve, 6 Air reservoir, 7 Drain valve, 8 Relay valve, 9 Load-sensing valve,
10 Diaphragm actuator, 11 Combination brake cylinder, 12 To brake.*

VA Front axle, HA Rear axle.

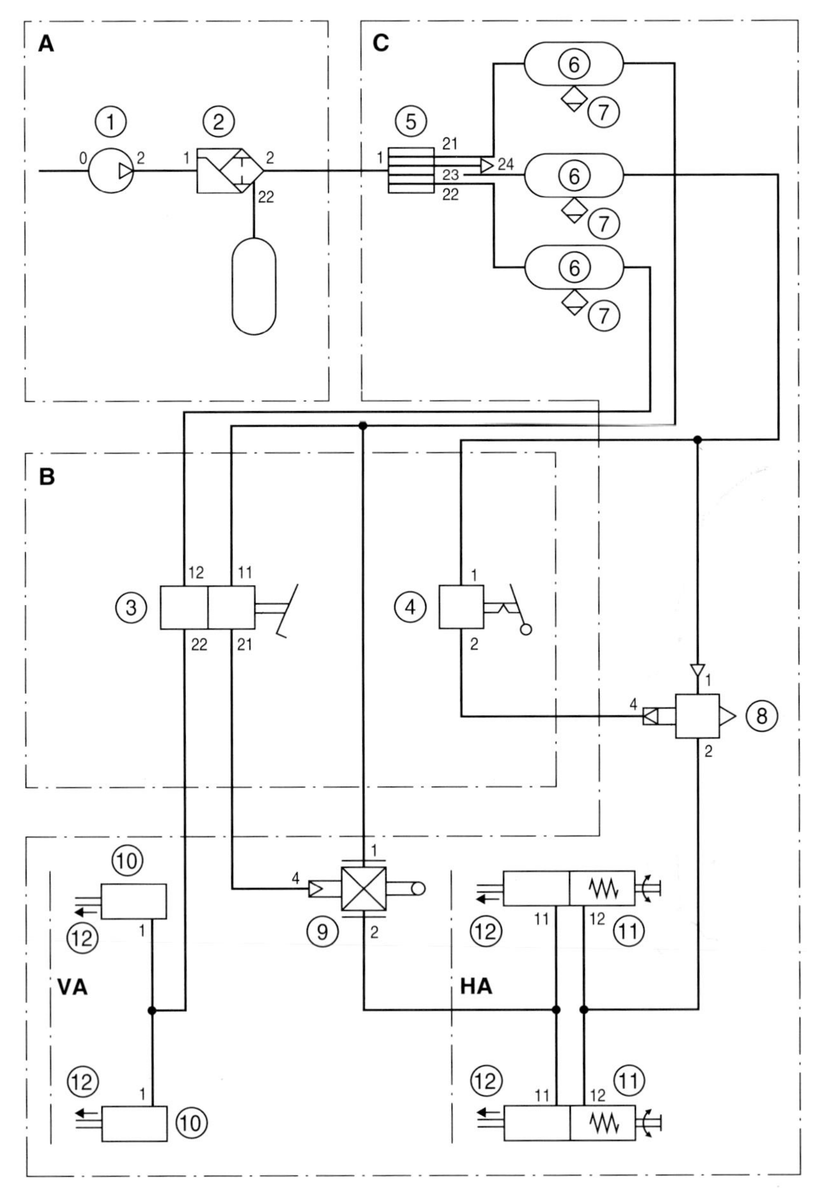

pads against the drums/discs as they turn along with the wheels. This friction transforms the moving vehicle's kinetic energy into thermal energy.

Drum brakes

This design employs friction forces generated on the inner surface of the brake drum. On the conventional internal-shoe brake, a self-energization effect amplifies the effective braking force to a higher level than would correspond to the direct tensioning force being applied through the system.

The self-energization effect occurs when the friction acting against the leading shoe generates torque at the shoe's fulcrum. This torque supplements the normal tensioning force at the brake drum. A corresponding but inverted torque is generated at the fulcrum on the trailing shoe; this force counteracts the tensioning force being applied through the system, producing a self-reduction effect.

This undesirable phenomenon is restricted to the simplex brake (Fig. 3). The duplex and servo designs (Figures 4

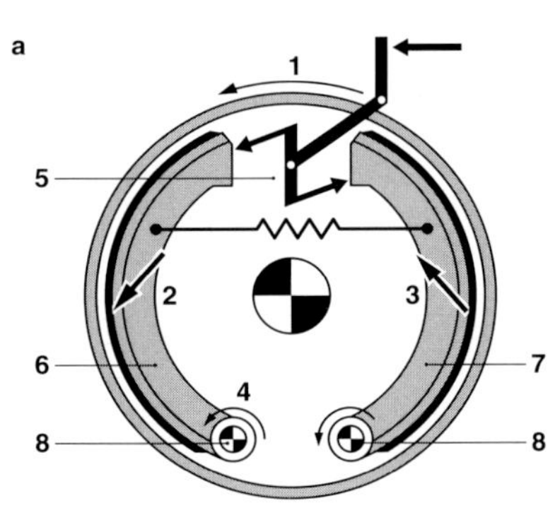

Fig. 3: Drum brake (Simplex brake),
a) with S-cam, b) with wedge
One leading brake shoe, low self-energization.
1 Direction of rotation of brake drum,
2 Self-energization, 3 Self-reduction, 4 Torque,
5 (a) S-cam or 5 (b) Wedge,
6 Leading shoe, 7 Trailing shoe, 8 Support point.

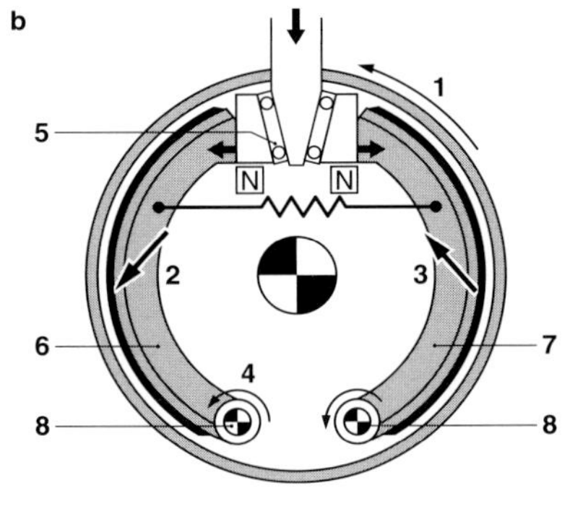

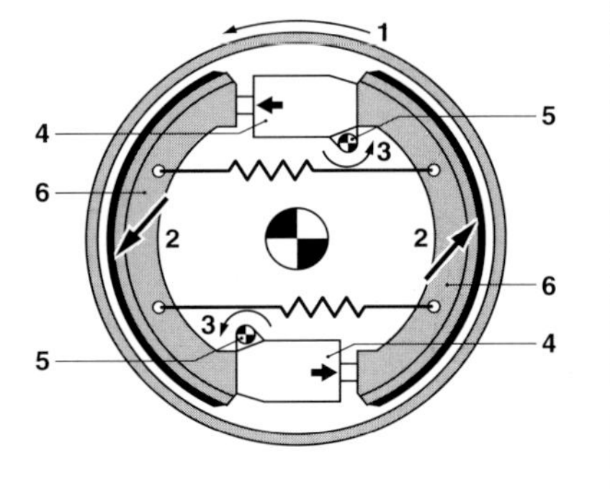

Fig. 4: Drum brake (duplex brake)
Two leading brake shoes, increased self-energization.
1 Direction of rotation of brake drum, 2 Self-energization, 3 Torque, 4 Wheel-brake cylinder 5 Support point, 6 Leading brake shoe.

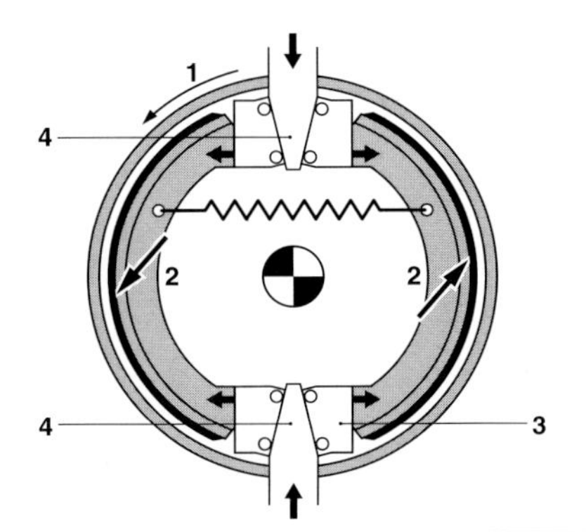

Fig. 5: Brake drum (duo-duplex brake)
Two leading brake shoes, floating brake application, self-energization increased even further.
1 Direction of rotation of brake drum,
2 Self-energization, 3 Wedge-type shoe,
4 Wedge.

and 5) feature self-energization at both brake shoes.

Disc brakes

In disc-brake systems (Fig. 6), the friction forces are generated against the surface of one or several discs or rotors.

Advantages:

The exposed friction surfaces provide much better thermal dissipation than is available with drum brakes. The discs are cleansed of brake dust automatically during operation, leading to more uniform braking. Because disc brakes do not display self-energization, the effect that fluctuations in the friction coefficient between pads and discs exercise on the braking process is minimal. This factor is of particular importance in maintaining braking balance between the left and right sides of the vehicle.

Disadvantages:

The disc's exposed friction surfaces are more sensitive to contamination and moisture than the surfaces on drum brakes. The absence of self-energization also means that higher contact forces are required.

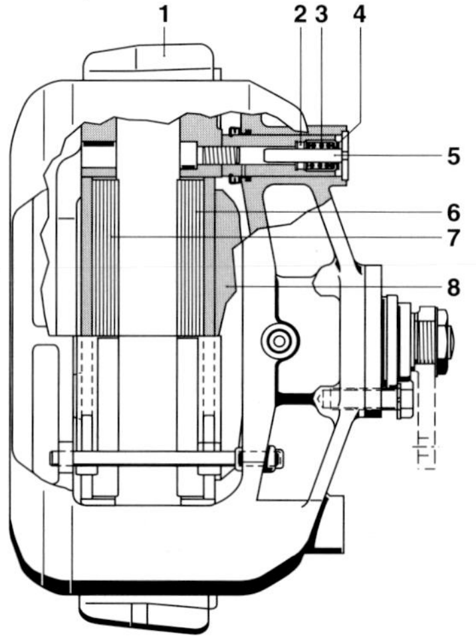

Automatic brake adjusters

Brake-lining (or pad) wear increases the clearance between brake lining (or pad) and drum (or disc). This leads to higher effective shoe or pad travel. Incorrect clearance leads to excessive travel for the wheel-brake cylinder piston; and in extreme cases, this can lead to a complete loss of braking force.

In the case of the brake drum, automatic rod adjusters reset the mechanism for the correct clearance when the brakes are released.

The distance through which the piston must travel in the wheel-brake cylinder during braking – equal to the total clearance – can be divided into three sections:

– the clearance between friction lining and brake drum or disc inherent in the particular design,
– additional clearance resulting from lining/pad wear,
– clearance stemming from compliance in the brake drum or disc and the linings, or originating in the force-transfer process between wheel-brake cylinder and wheel brake (compliance travel).

The automatic adjuster provides the required compensatory adjustments.

Retarders

Retarders, like friction brakes, can be employed to reduce a vehicle's speed; however, they differ in being unsuitable for actually halting the vehicle. Even under extended braking, the unit can still dissipate the heat generated during operation. This means that retarders can be regarded as virtually wear-free.

In contrast to the friction brake, the retarder is suitable for protracted use and makes an especially important safety contribution during descents on extended gradients by relieving the friction brakes, which remain fully-operational for emergency use.

Fig. 6: Pneumatically actuated disc brake
No self-energization (principle of the brake).
1 Brake anchor plate,
2 Grip ring,
3 Compression spring,
4 Screw,
5 Bolt,
6 and 7 Brake linings,
8 Pressure plate.

Exhaust brake (engine brake)
The engine brake operates by increasing the engine's internal resistance to motion. Engaging the engine brake interrupts the flow of fuel to the engine, while a rotary or pivot-type valve installed in the exhaust line prevents the air drawn in through the engine's intake tract from flowing out through the exhaust system. This generates an air cushion in the cylinder, and brakes the piston in the compression and exhaust strokes. The engine brake has no provision for graduated operation.

Hydrodynamic retarder
(Hydraulic retarder, Fig. 7)
This retarder design comprises a rotating stator, the braking rotor (3), and a stationary, or braking stator (1) which are located opposite each other in a common housing. The driveshaft used to power the vehicle's wheels also provides the mechanical link between the braking rotor and the drivetrain. When this brake is applied, the chambers in the braking rotor and stator fill with fluid. The braking rotor accelerates the fluid while the stator retards it, and the kinetic energy present in mechanical form at the braking rotor is transferred to the fluid.
This decelerates the braking rotor, and this braking effect is transferred through the driveshaft to the vehicle's wheels. The friction in the fluid is converted to heat, which is then absorbed by the engine's coolant. The fluid charge can be varied for infinitely variable control of braking effect.

Electrodynamic retarder
(Eddy-current brake, Fig. 8)
This device consists of an air-cooled soft-iron disc rotating in a variable electromagnetic field generated by the battery.
The resulting eddy currents retard the disc and automatically brake the vehicle's wheels. The braking effect is infinitely variable.

Fig. 7: **Hydrodynamic retarder (hydraulic retarder)**
1 Braking stator,
2 Drive shaft,
3 Braking rotor,
4 Mounting flange,
5 Heat exchanger (oil/water)

Fig. 8: **Electrodynamic retarder (eddy-current brake)**
1 Star-shaped bracket, 2 Rotor (transmission side), 3 Shim rings (air-gap setting), 4 Stator with coils, 5 Intermediate flange, 6 Rotor (rear-axle side), 7 Gearbox cover, 8 Gearbox output shaft, 9 Air gap.

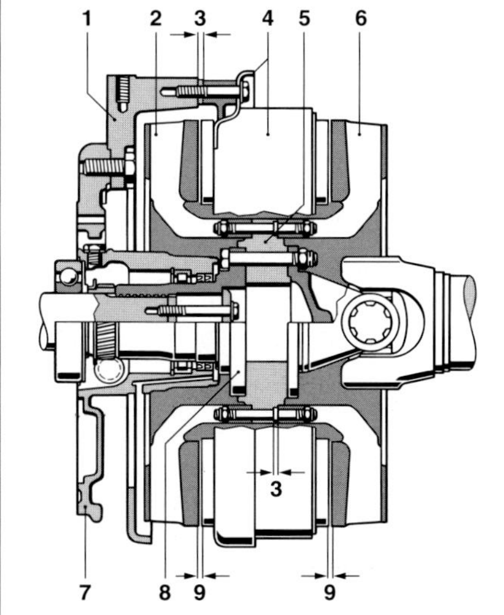

Brake systems

Brake systems can be classified according to the following criteria:
– Design features, and
– Operating concepts.

Designs

Service-brake system

The service brakes ("foot brake") can be used to reduce the vehicle's speed or, alternatively, to maintain it at a constant level on downgrades. They can also bring the vehicle to a halt. This is the system employed during normal vehicle operation. The pedal allows precise, variable control of a system which operates on all vehicle wheels.

Single-circuit brake system (Fig. 1)
The most basic type of system employs a single transmission (or force-transfer) device – generally referred to as the brake circuit – to relay braking energy to all of the vehicle's wheels. This is the single-circuit brake system.
A defect in any part of this circuit will lead to failure of the entire brake system.
This manual focuses on dual-circuit brake systems, as this is the only configuration approved in the EU.

Dual-circuit brake system (Fig. 2)
To enhance safety, the brake system's transmission device incorporates two circuits. On commercial vehicles the usual practice is to assign one brake circuit to the wheel-brake cylinders on the front axle, and the second to those on the rear axle. Should one of the brake circuits on the dual-circuit system fail, the second, intact circuit still remains operational.
Assuming that it is sufficiently effective, the surviving section of the dual-circuit brake system can serve as the secondary brake system.

Secondary brake system

In the event of failure in the service-brake system, the secondary brake system must be capable of assuming its functions, viz., to reduce the vehicle's speed, to maintain it at a constant level on downgrades, and to bring the vehicle to a halt. The secondary brake system may operate at a lower level of effectiveness, and need not necessarily consist of a third, independent system (augmenting the service and parking brakes); it may also consist of the remaining, intact circuit in a dual-circuit brake system or of a parking brake designed for graduated operation.

Parking-brake system

The parking brakes ("hand brake") assume the third function assigned to the

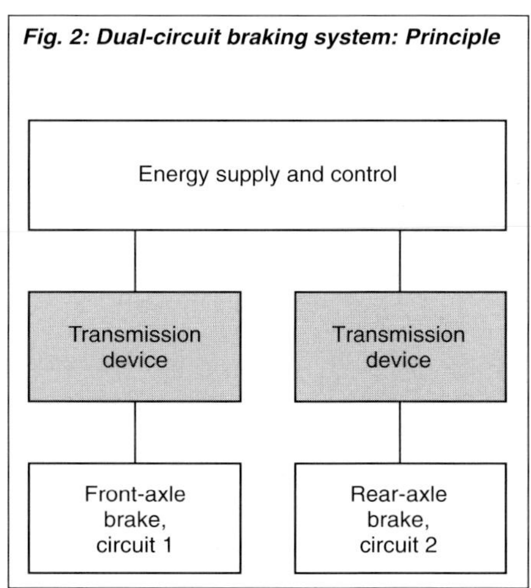

Fig. 1: Single-circuit braking system: Principle

Fig. 2: Dual-circuit braking system: Principle

brake equipment. The parking brake must maintain the vehicle in a stationary state, remaining effective on slopes and in the absence of the driver.

To ensure safety, the parking brakes must remain fully operational in the event of a pneumatic or hydraulic system failure.

This requirement is met by using a continuous mechanical system to connect the energy supply and the wheel brake, e.g., mechanical linkage or a cable running between the hand-brake lever and the wheel brake. An alternative is to employ a compressed-air cylinder connected to the wheel brake.

The parking brakes on multi-vehicle combinations are controlled from the driver's seat – usually with a hand lever – with a trailer-mounted handbrake lever available for separate control of the trailer brakes. This system operates on the wheels on one or several axles.

Parking brakes employed as a secondary brake system must be designed for precisely-controlled, graduated operation.

Brake systems on combination vehicles

Tractors (towing vehicles) in combinations (consisting of the tractor together with the independent trailer or semi-trailer) are equipped with a supplementary unit to supply energy to the trailer brakes and to control them.

This unit is required to transmit from the tractor the braking energy and the control pulses required for braking the trailer.

As the designations imply, single-line brake systems employ a single line for these transmission functions, while the dual-line configuration uses two lines.

Overrun brakes (not described) are approved for use on trailers with an AGVW of up to 3.5 metric tons. The energy required for braking is generated as the trailer tries to run-on and approach the tractor when the tractor brakes are applied.

Single-line brake system

The single-line brake system employs one line for two functions: It charges the trailer's energy-accumulator unit (compressed-air tank or reservoir) – providing the braking energy. It also controls the trailer brakes via the trailer brake valve, which responds to pressure drops in the control line. Single-line brake systems are no longer approved for use in the member nations of the EU, and the following text dispenses with further descriptions of this type of system to concentrate on the dual-line brake system.

Dual-line brake system

The dual-line brake system employs two pneumatic lines – the "supply" line and the "brake" line – to link the individual vehicles of the combination. The brake line operates by transmitting increasing pressures for braking. As the supply line can furnish a continuous flow of compressed air to the trailer reservoir, the energy supply for the trailer brakes is inexhaustible. The coupling heads (gladhands) on the supply and brake lines are specially designed to ensure that each is connected to the proper fitting. Should the trailer inadvertently become disconnected from the tractor, the resulting pressure loss in the supply line will lead to automatic application of the trailer brakes.

Operating concepts

Depending upon the configuration of the transmission device, a distinction is drawn between:
- Muscular-energy brake systems,
- Energy-assisted (power-assisted) brake systems, and
- Non-muscular energy (power) brake systems.

Muscular-energy brake system

This type of system is fitted in passenger cars and light commercial vehicles.

Either a mechanical (e.g., linkage rods or Bowden cable) or hydraulic mechanism transfers the muscular energy applied at the pedal or hand lever directly to the wheel brakes.

Power-assisted brake system

This type of system is also installed in passenger cars and light commercial vehicles. A brake servo-unit employs supplementary force supplied by compressed air, vacuum or hydraulic fluid to augment the muscular force. This supplementary force is applied to the mechanically or hydraulically actuated wheel brakes.

Because the linkage between the control mechanism and the brakes is continuous, the driver retains the ability to brake the vehicle using muscular force alone, albeit with reduced braking effectiveness and/or higher levels of force at the control mechanism.

Power-brake system

Medium-heavy and heavy-duty commercial vehicles are equipped with brake systems that operate exclusively with application forces generated using extraneous (non-muscular) sources such as compressed air, vacuum pressure or hydraulic fluid.

The driver's muscular force exercises nothing more than a control function, and cannot be used to generate braking forces should the energy source become unavailable.

Brake systems with hydraulic energy as primary or boost medium

This type of system employs hydraulic energy (hydroscopic pressure) as well as a hydraulic transmission device. The hydraulic fluid is stored in energy reservoirs (hydraulic accumulators) containing a compressed gas (usually nitrogen). A diaphragm (bladder diaphragm) or a piston with rubber seal (piston accumulator) insulates the gas from the fluid. A hydraulic pump generates the hydraulic pressure required to maintain a state of equilibrium between gas and hydraulic fluid, while a pressure regulator cycles the hydraulic pump to idle whenever the system reaches its maximum pressure. The actuating device in the non-muscular energy (power) brake system is the brake valve, a function assumed by the

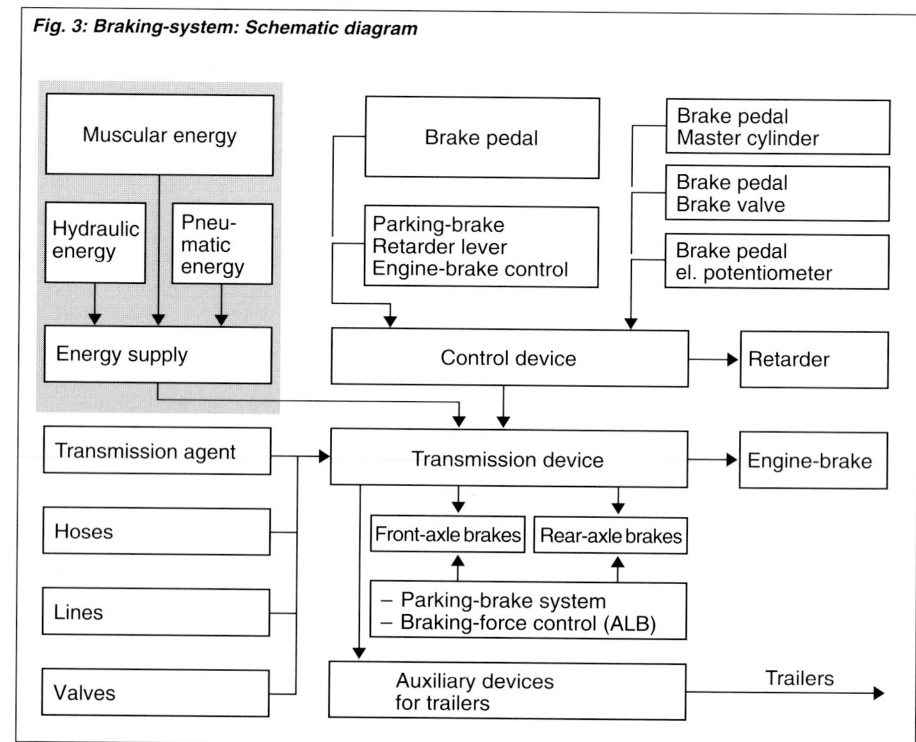

Fig. 3: Braking-system: Schematic diagram

brake booster in the energy-assisted (power-assisted) system.

Advantages:
Hydraulic fluid differs from pneumatic media in being virtually incompressible, maintaining a constant volume regardless of the pressure being exerted on it. The system can generate high pressures, as small surface areas can be employed to transfer high levels of force; system components can be more compact than those used in compressed-air systems.

Disadvantages:
A system leak will lead to loss of hydraulic fluid and, ultimately, of braking effectiveness. Trailers are rarely equipped with hydraulic systems, as it is difficult to provide a satisfactory connection for transmitting hydraulic fluid from the tractor to the trailer. The difficulties include the tendency to leak as well as problems arising due to potential entry of dirt and air into the system.

Power-assisted brake systems
with pneumatic energy
as the primary or boost medium
This type of system employs pneumatic energy in the form of vacuum or compressed air. This energy is applied in a brake booster or in valve-controlled wheel-brake cylinders.

Vacuum-brake systems
On spark-ignition engines, the vacuum is obtained through a line leading to the intake manifold; on diesel engines it is generated by a vacuum pump. The pressure obtained when connected to the manifold varies according to the suction effect generated by the engine. Approximate values are 1.0 bar at full-throttle, 0.5 bar at idle and 0.7 bar after abrupt transitions to trailing throttle at high engine speeds. The level of vacuum available from a vacuum pump remains relatively constant at 0.7...0.9 bar, regardless of engine speed. Pressures in vacuum-brake systems are specified in absolute or negative pressure.

Advantages:
Because only limited amounts of atmospheric humidity can enter the system, there is no danger of ice formation in winter. Contemporary vacuum-brake systems generally feature dual chambers with two series-connected pistons; and because only limited air flow is required for operation, these provide rapid response despite the small pressure differential.

Disadvantages:
This same low pressure differential with respect to the atmosphere means that large units are necessary to generate the required braking force.

Compressed-air brake systems
Air is compressed in a single or multi-stage, engine-powered, water or air-cooled compressor designed to draw in air whenever the engine is running. A pressure regulator cycles the compressor between pressure supply and idle. At idle the compressor discharges the air back into the atmosphere.
The pressures specified for compressed-air brake systems are positive.

The operative distinction is between
low-pressure ($p_e < 10$ bar), and
high-pressure systems ($p_e > 10$ bar).

Advantages:
The greater difference between system and atmospheric pressures makes it possible to use smaller components than with a vacuum system, as small surfaces areas can be employed to generate high levels of braking force when higher forces are applied.

Disadvantages:
Condensation forms when humid air is compressed. Thus special equipment, such as an air dryer, is required to prevent condensed moisture from entering the system.

Braking action

Calculations

The following section is designed to provide a source of in-depth information on the theoretical concepts and the underlying relationships used to describe brake-system operation.

Symbols

a	Instantaneous braking deceleration in m/s²
a_m	Mean braking deceleration in m/s² over total braking time t_B
a_{max}	Maximum deceleration in m/s²
F	Braking force in N
g	Acceleration due to gravity (9.81m/s²)
G	Weight (force) in N
G_z	Permissible total weight force of the laden vehicle in N
m	Vehicle weight in kg
m_z	Approved gross vehicle weight, kg
p	Brake pressure in bar
p_b	Design pressure in bar
s_o	Pre-braking distance in m
s_B	Braking distance in m
s_{tot}	Total braking distance in m
t_0	Reaction time in s
t_a	Brake response time
t_B	Braking time in s
t_{tot}	Total braking time in s
t_s	Threshold time in s
t_v	Period of full braking effect in s
v	Instantaneous velocity in m/s
v_1	Velocity at start of braking in m/s
v_2	Velocity at end of braking in m/s
z	Instantaneous retardation
z_{max}	Maximum retardation
ε	Braking efficiency
μ_{HF}	Static coefficient of friction (braking-force coefficient)

Braking deceleration

The braking deceleration a is a mathematical operand for braking action. It is the opposite to acceleration and defines the reduction in vehicle speed over time. As shown in Fig. 1, it also varies continuously throughout the total braking time t_B. Due to system lag (play in the system), there is no braking deceleration during the brake response time t_a. Braking deceleration climbs to its maximum value of a_{max} (maximum deceleration) during the threshold time t_s; the system then maintains this rate until the braking process is completed. Calculations for the total effective braking time t_B are thus based on the mean braking deceleration $a_m = (v_1 - v_2)/t_B$. If the vehicle – like the one in Figure 1 – is stationary ($v_2 = 0$) upon termination of the braking process, then mean braking deceleration is calculated using the following equation (1):

$$a_m = \frac{v_1 - v_2}{t_B} = \frac{v_1 - 0}{t_B} - \frac{v_1}{t_B}$$

The braking time t_B is required for calculations of the mean braking deceleration. To simplify matters, calculations are based on the operative assumption that the first half of threshold time t_s is characterized by no braking deceleration, while maximum deceleration a_{max} is present throughout the second half of t_s. Referring to Figure 1:

$$t_B = t_a + \frac{t_s}{2} + \frac{v_1}{a_{max}}$$

In the example shown in Figure 1, the effective braking time :

$$t_B = 0.2\,s + \frac{0.4\,s}{2} + \frac{19.4\,m/s}{6.2\,m/s^2} = 3.5\,s$$

The equation (1) can now be used to obtain the mean braking deceleration:

$$a_m = \frac{19.4\,m/s}{3.5\,s} = 5.5\,m/s^2$$

Braking retardation

The braking factor z provides a second index of braking effectiveness. It represents the ratio between braking force F and vehicular weight force G:

$$z = \frac{F}{G}$$

The relationship between braking factor z and braking deceleration a is defined thus:

$$a = z \cdot g$$

The retardation z is usually measured on a roller-type brake dynamometer. The ratio of braking force F (just before the wheels lock), to the specified vehicle weight force G can be used to calculate the maximum retardation z_{max}.

On commercial vehicles the braking retardation is defined as the ratio of braking force F to the approved gross weight force G_z of the laden vehicle. To test the brake system, it is not necessary to load the vehicle to its full capacity before driving it onto the brake dynamometer. Adequate results for compressed-air brake systems – on partially loaded or unladen vehicles – are available using an extrapolation based on the brake pressure p required for the empirically derived braking force F:

$$z = \frac{F}{G_z} \cdot \frac{p_b - 0.4\ \text{bar}}{p - 0.4\ \text{bar}}$$

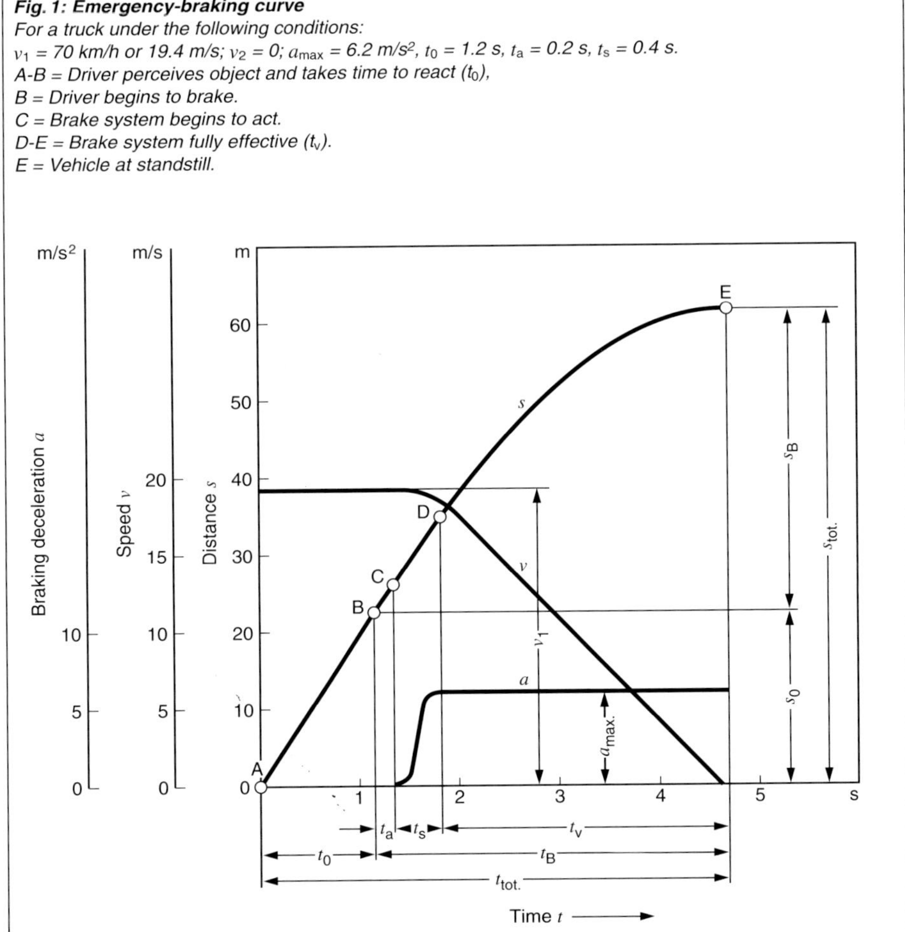

Fig. 1: Emergency-braking curve
For a truck under the following conditions:
v_1 = 70 km/h or 19.4 m/s; v_2 = 0; a_{max} = 6.2 m/s², t_0 = 1.2 s, t_a = 0.2 s, t_s = 0.4 s.
A-B = Driver perceives object and takes time to react (t_0),
B = Driver begins to brake.
C = Brake system begins to act.
D-E = Brake system fully effective (t_v).
E = Vehicle at standstill.

Here the response pressure required to activate the brake system is 0.4 bar.

Example:
A truck with an approved gross vehicle weight m_z = 17 metric tons (17000 kg) has an approved gross weight force

$$G_z = 17\,000\,kg \cdot 9.81\frac{m}{s^2} \approx 170\,000\,N$$

and achieves a measured braking force of F = 53500 N on the brake dynamometer, obtained at brake pressure p = 3.2 bar. The design pressure of the brake system is p_b = 6.0 bar. These data furnish the basis for determining the maximum braking retardation at the approved gross vehicle weight:

$$z_{max} = \frac{53\,500\,N}{170\,000\,N} \cdot \frac{6.0\,bar - 0.4\,bar}{3.2\,bar - 0.4\,bar} = 0.63$$

As a reference, the EU nations prescribe the following minimum retardation rates for passenger buses:

Service brakes	0.50
Secondary brakes	0.25 and
Parking brakes	0.18.

Total braking time
The total braking time t_{tot} extends from the point at which the obstacle is perceived until the vehicle stops (Fig. 1). It is calculated as follows:

$$t_{tot} = t_0 + t_s + \frac{t_s}{2} + \frac{v_1}{a_{max}}$$

In the example illustrated in Figure 1 the total braking time is:

$$t_{tot} = 1.2\,s + 0.2\,s + \frac{0.4\,s}{2} + \frac{19.4\,m/s}{6.2\,m/s^2} = 4.7\,s$$

Total braking distance
To facilitate calculations, the equation for total braking distance s_{tot} – representing the total distance traversed between the two points defined in the section above – is based on the following assumptions: the velocity to the end of the first half of

the threshold time t_s will be v_1; for the second half of t_s the mean velocity is $(v_1 + v_2)/2$, with v_2 = 0, thus $v_1/2$. This results in:

$$s_{tot} = v_1 \left(t_0 + t_a + \frac{t_s}{2}\right) + \frac{v_1}{2}\left(t_B - t_a + \frac{t_s}{2}\right)$$

providing:

$$s_{tot} = \frac{v_1}{2}\left(2t_0 + t_B + t_a + \frac{t_s}{2}\right)$$

In the example shown in Figure 1, the total braking distance is:

$$s_{tot} = \frac{19.4\,m/s}{2}\left(2 \cdot 1.2 + 3.5 + 0.2 + \frac{0.4}{2}\right)s = 61.1\,m$$

Static coefficient of friction (brake-force coefficient)
The braking force F exerted at the wheel's circumference must be transmitted to the road surface through frictional force. The ratio of the braking force corresponding to maximum transfer of frictional force, relative to the effective weight force at the wheel, is defined as the static coefficient of friction μ_{HF} (or traction or adhesion coefficient, describing the traction available between tire and road surface):

$$\mu_{HF} = \frac{F}{G}$$

The numerical expression of the coefficient of friction μ_{HF} corresponds to the maximum theoretically available (ideal) braking retardation. It varies according to:
- Tire condition,
- Vehicle speed, and
- Road surface adhesion.

Should the braking force continue to increase beyond the point of maximum frictional transfer, the result will be higher levels of braking slip, with potential wheel lock; this is generally accompanied by a reduction in frictional forces.
Wheel lock, which poses an extreme threat to vehicle stability, can be effectively prevented by an antilock braking system (ABS).

Road condition	Static coefficient of friction μ_{HF}
Dry	0.8...1
Wet	0.2...0.65
Ice	0.05...0.1

Braking efficiency
The braking efficiency ε is the ratio between the actual maximum retardation factor z_{max} and ideal retardation (corresponding to μ_{HF}):

$$\varepsilon = \frac{z_{max}}{\mu_{HF}}$$

This ratio is understood to describe the ultimate efficiency with which a given brake system operates.

Brake specifications
(under emergency braking conditions and at various speeds)

Braking conditions			Vehicle speed v_1 at start of braking, **20 km/h** (5.6 m/s)		
Static coefficient of friction	Max. retardation	Max. braking deceleration	Mean braking deceleration	Total braking time	Braking distance
μ_{HF}	z_{max}	a_{max}	a_m	t_B	s_B
–	–	m/s^2	m/s^2	s	m
0.22	0.20	1.96	1.72	3.26	10.2
0.44	0.40	3.92	3.06	1.83	6.2
0.67	0.60	5.89	4.15	1.35	4.9
0.89	0.80	7.85	5.05	1.11	4.2

Braking conditions			Vehicle speed v_1 at start of braking, **50 km/h** (14 m/s)		
Static coefficient of friction	Max. retardation	Max. braking deceleration	Mean braking deceleration	Total breking time	Braking distance
μ_{HF}	z_{max}	a_{max}	a_m	t_B	s_B
–	–	m/s^2	m/s^2	s	m
0.22	0.20	1.96	1.86	7.54	55.6
0.44	0.40	3.92	3.53	3.97	30.6
0.67	0.60	5.89	5.04	2.78	22.3
0.89	0.80	7.85	6.42	2.18	18.1

Braking conditions			Vehicle speed v_1 at start of braking, **80 km/h** (22.4 m/s)		
Static coefficient of friction	Max. retardation	Max. braking deceleration	Mean braking deceleration	Total braking time	Braking distance
μ_{HF}	z_{max}	a_{max}	a_m	t_B	s_B
–	–	m/s^2	m/s^2	s	m
0.22	0.20	1.96	1.89	11.62	132.2
0.44	0.40	3.92	3.66	6.01	70.5
0.67	0.60	5.89	5.31	4.14	49.9
0.89	0.80	7.85	6.87	3.20	39.6

ABS antilock braking system

Why ABS?

The antilock braking system – or ABS – from Bosch greatly enhances braking safety in commercial vehicles. On both normal and low-traction road surfaces, it prevents the wheels from locking in response to excess brake pressure. When the brakes are applied on a road surface characterized by an asymmetrical left/right traction pattern (for instance, left wheels on dry asphalt, right wheels on ice), the resulting brake forces vary. This leads to yaw (torque in the form of a tendency for the vehicle to spin around its verticle axis).

A yawing moment build-up delay (GMA) device makes it easier to keep the vehicle on course.

On ABS-equipped vehicles this means:
– The vehicle retains directional stability,
– Vehicle steerability is maintained,
– No breakaway tendency of the trailer/rear section on combination and articulated vehicles,
– Optimum deceleration rates are obtained during panic braking, and
– Less driver countersteer is required on road surfaces providing uneven left/right traction.

Control concept

Figure 1 illustrates how the ABS components are integrated within the compressed-air service-brake system on a two-axle commercial vehicle.

ABS uses pulse rings (2) attached to the insides of the wheels to monitor their individual rotational speeds. As they rotate, the pulse rings generate pulses in the stationary wheel-speed sensors (1) which is proportional to the wheel's rotational speed. The computer in the electronic control unit (ECU, 3) uses the change in wheel-speed that accompanies braking to determine the wheels' deceleration (negative acceleration $-a$), acceleration a and brake slip λ. These provide the basis for calculations to determine the brake pressure that will supply maximum retardation without leading to locked wheels.

The ECU (3) transmits control pulses to the solenoid units at the pressure-control valves (4). This is where the data from the ECU regulate the pressure from the service-brake valve (6) to ensure that the forces generated in the wheel-brake cylinders (10, 7) brake the wheels with maximum efficiency and without wheel lock.

The curves in Figure 2 show, for various types of road surface, how the friction coefficient, and with it the braking action,

---

Fig.1: ABS/ASR for 2-axle commercial vehicle with ASR brake controller
1 Wheel-speed sensor, 2 Sensor ring, 3 ABS/ASR electronic control unit, 4 Pressure-control valve, 5 Shuttle valve, 6 Service-brake valve, 7 Combination brake cylinder, 8 Air reservoir, 9 Drain valve, 10 Diaphragm actuator, 11 ASR solenoid valve, 12 Relay valve.

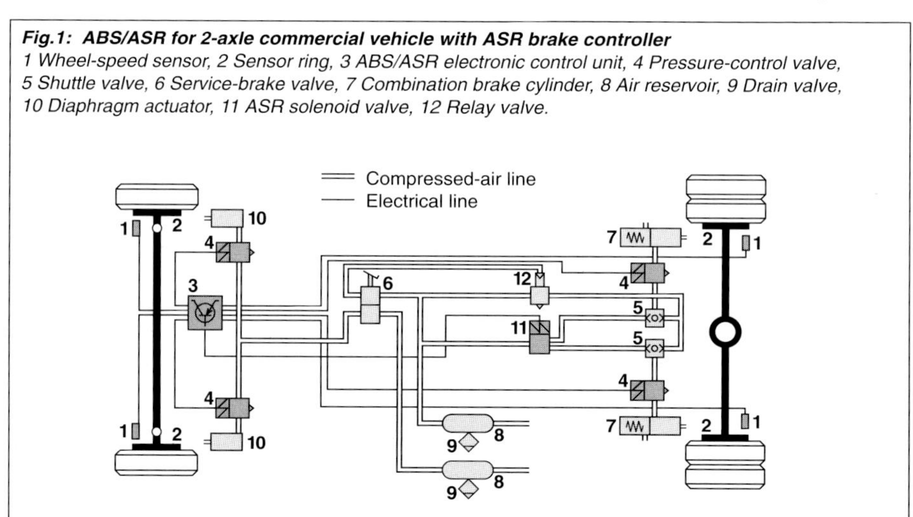

increase to a maximum as a function of brake pressure.

On a vehicle without ABS, the brake pressure can be increased beyond these maxima so that overbraking immediately takes place. The resulting tire deformation means that the contact patch between the slipping tire and the road surface increases to such an extent that the coefficient of friction starts to drop and brake slip

increases. Finally, the wheel locks (Point B).

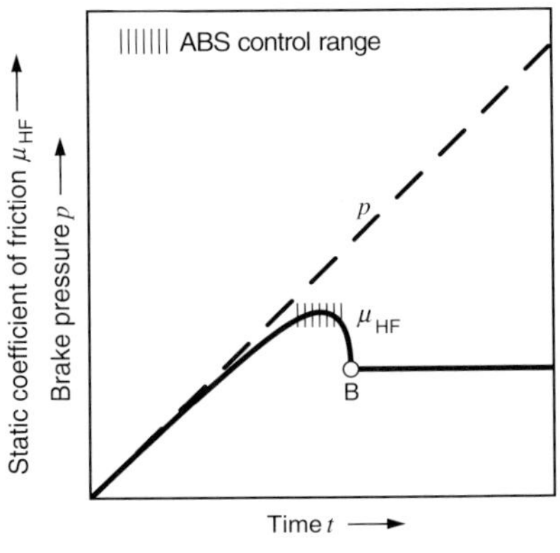

Explanation:

Frictional processes can be divided into static and and sliding friction. The static friction for solid masses is greater than the sliding friction. As this implies, there are conditions under which the coefficient of friction at a rotating rubber tire is higher than when the wheel is locked. Sliding processes also occur as the rubber tire rotates. This condition is referred to as slip. How quickly this point is reached can be seen from the steep drop in the friction curve in Figure 3.

$$\lambda = \frac{(v_F - v_R)}{v_F} \cdot 100\,\%$$

illustrates the degree to which the wheel's peripheral velocity v_R lags behind vehicle speed v_F.

Whether or not the vehicle is equipped with ABS, most braking remains in the stable area to the left of the ABS control range. ABS only becomes active in response to overbraking. It uses closed-loop control to prevent the brake pressure from entering the instable range (to the right of the control range) characterized by high rates of brake slip and the attendant danger of wheel lock.

The system also provides optimum braking performance by making maximum use of the available friction between tire and road surface. During cornering, the available traction is divided between decelerative and lateral components. The system also prevents locking due to excessive braking under these conditions, provided that the maximum cornering speed is not exceeded.

In summary, ABS provides maximum utilization of the available braking forces within the technical boundaries imposed by physical limits.

This applies to braking distances as well as to cornering speed. Even with ABS, the driver must continue to maintain a driving style apprropriate to the road-surface, climatic and traffic conditions.

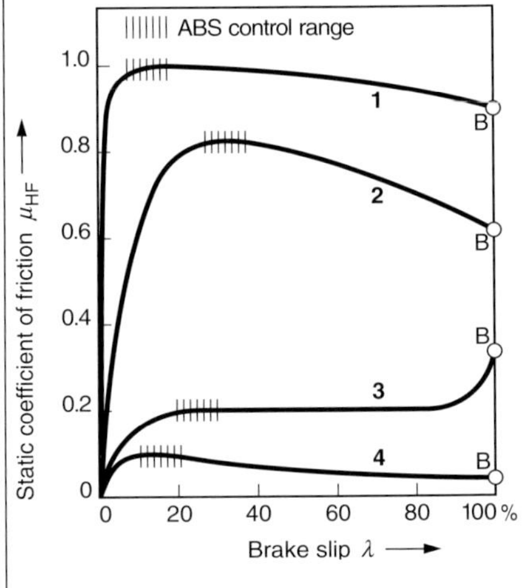

Fig. 2: Static coefficient of friction μ_{HF}
As a function of brake slip λ.
1 Radial-ply tires on dry concrete.
2 Diagonal-ply winter tires on wet asphalt.
3 Radial-ply tires on loose snow. 4 Radial-ply tires on wet black ice. 5 Start of wheel lock.

Fig. 3 : Time curve of static friction with increasing brake pressure
B Start of lock-up.

Components

The ABS antilock braking system consists of three subassemblies. The actual number of the particular subassemblies which are fitted depends upon vehicle type and the specific closed-loop control strategy:

– Wheel-speed sensor with pulse ring,
– Electronic control unit (ECU), and
– Pressure-control valve (solenoid relay valve in trailers).

Wheel-speed sensor with pulse ring

As the wheel turns, the pulse ring mounted on the hub (Fig. 4) generates AC voltage at the wheel-speed sensor. This voltage frequency is directly proportional to the wheel's rotational speed.The wheel-speed sensor is held in place by a spring sleeve. During initial installation it is positioned directly against the pulse ring; in the course of the ensuing vehicle operation, wheel-bearing float and component compliance automatically provide the required clearance to the wheel-speed sensor.

A wheel-speed sensor with pulse ring is installed at each wheel (or each pair on dual-wheel axles). This layout makes it possible to apply closed-loop control of brake pressure at each individual (dual) wheel. The result: Maximum stability and minimum braking distances.

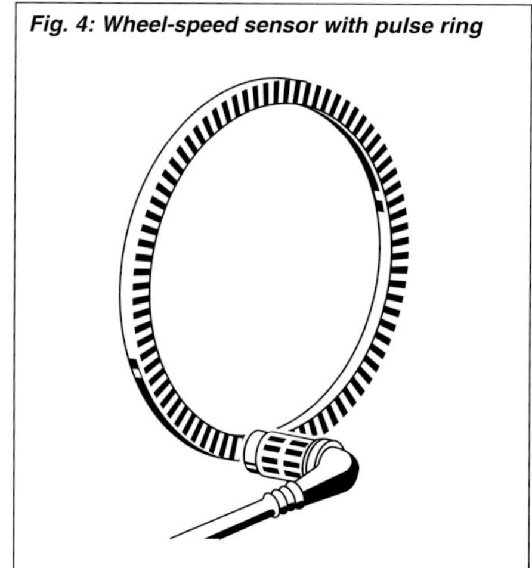
Fig. 4: Wheel-speed sensor with pulse ring

Electronic control unit (ECU)

The electronic control unit's input stages, which are based on digital circuit technology, convert the incoming sinusoidal signals from the wheel-speed sensors into square-wave signals. Microcomputers designed for system redundancy use the frequency of the square-wave signals as the basis for calculating the wheels' rotational speeds and acceleration rates (Fig. 5).

The microcomputers also use the speeds of two diagonally opposed wheels to calculate a reference speed for the vehicle. The brake slip at each wheel is derived by comparing this reference speed with the speeds of the individual wheels. The "wheel acceleration" and "wheel slip" signals serve to alert the ECU to any locking tendency. The microcomputers respond to such an alert by sending a signal through the ECU output stages to trigger the pressure-control valve solenoids (refer to following section). These in turn modulate the brake pressure in the individual wheel-brake cylinders.

The ECU also incorporates a number of features designed to support error recognition for the entire ABS system (wheel-speed sensors, the ECU itself, pressure-control valves, wiring harness). The ECU reacts to a recognized defect or error by switching off the malfunctioning part of the system, and at the same time storing a code containing the error path. The conventional service-brake system remains fully operational.

On combination vehicles, each component – tractor and trailer(s) – is equipped with its own ABS. The tractor is also equipped with an auxiliary unit to monitor trailer status. By means of warning and indicator lamps, this unit informs the driver of the equipment level and the current status of the trailer's ABS:

– When the indicator lamp comes on, this signals that the trailer is not equipped with ABS.
– Simultaneous illumination of the warning and indicator lamps means "Trailer ABS malfunction."

Pressure-control valve

The pressure-control valve applies individual levels of brake pressure to the wheel-brake cylinders.
It comprises two solenoid valves which control the diaphram valves (Fig. 6).

Legal requirements

The safety benefits provided by ABS have received official recognition:
Since 1 October, 1991, EU braking systems Directive 71/320 EEC has made ABS mandatory equipment for the initial registration of the following classes of vehicles:
– Buses of more than 12 metric tons,
– Trucks (for trailer operation) and tractor/semitrailer rigs of 16 metric tons and above, and
– Trailers with an AGVW of 10 metric tons or above.

Operation

During braking, the ABS installed on commercial vehicles fulfills two functions:

– It prevents the wheels from locking (controlled braking), and
– It reduces the tendency of the vehicle to yaw – rotate around its vertical axis – during braking on surfaces affording different traction levels on the left and right sides (yawing-moment limitation).

Controlled braking

During the braking process, the driver applies force to the brake pedal in order to transfer a specific level of brake pressure to the wheel-brake cylinders and thus decelerate the wheels. The response pattern displayed by the two wheels at the vehicle's rear axle during braking time t is illustrated in Figures 7 and 1:
Referring to Fig. 1, Page 88, when the driver applies the brakes, brake-pedal actuation starts at t_1. Compressed air flows from the service-brake valve (6) and through the relay valve (12) and open pressure-control valves (4) on its way to the brake cylinders (7). Pressure continues to increase until signals from one or both wheel-speed sensors indicate incipient lock to the ECU (3); this occurs at point t_2.
The ECU evaluates all incoming wheel-speed signals to derive a reference speed v_{Ref} which lies below the vehicle speed by a specific slip rate.
The following describes active ABS control at wheel 2 on a single axle.
At point t_2 the wheel's deceleration (negative acceleration $-a$) reaches a high enough level to trigger transmission of a closing signal to the pressure-control valve's (4) inlet valve. The brake cylinder (7) pressure is maintained at a constant level. At t_3 the slip rate stored in the logic circuits is exceeded, initiating a process in which the outlet valve at the pressure-

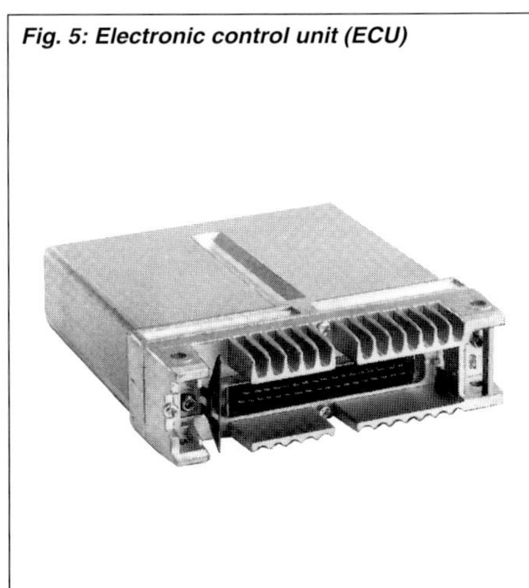

Fig. 5: Electronic control unit (ECU)

Fig. 6: Pressure-control valve

control valve (4) opens to release pressure from the brake cylinder (7). This pressure-release process reduces the wheel's deceleration rate (i.e., lower braking forces are applied to the wheel). This continues until the deceleration rate falls to point t_4, cancelling the pressure-release process at the brake cylinder (7). The outlet-valve closes. At point t_5 the wheel-speed then increases again (the deceleration $-a$ becomes acceleration $+a$); and continues to increase up to t_6, where the wheel speed equals the vehicle's reference speed. The vehicle brakes at optimal slip rates. At this point the brake pressure in the brake cylinder can be increased again.

To achieve the closest possible approximation to optimal slip during braking, instead of a continuous current the ECU applies pulses to trigger the inlet valve. The inlet valve closes when the t_2 deceleration is again reached at t_7, and a new closed-loop control cycle is initiated.

Yawing-moment build-up delay

The road surfaces encountered in the winter and in the wet are frequently non-uniform, with vast differences between the traction available at the lane's crown and at its shoulder. Such a road surface is said to be of the μ-split type. Under these conditions, applying the brakes will create a tendency for the vehicle to yaw around its vertical axis, regardless of whether or not it is equipped with ABS. If the left wheel 1 enjoys higher traction than the right wheel 2, then wheel 1 will tend to be braked much more than wheel 2 unless steps are taken to limit the yawing moment. The vehicle will attempt to rotate to the left around its vertical axis, and the driver is forced to respond by applying major steering corrections to maintain the vehicle in its lane.

It is possible to reduce the vehicle's tendency to rotate around its vertical axis by using a modified strategy for independent control at the front wheels. This stra-

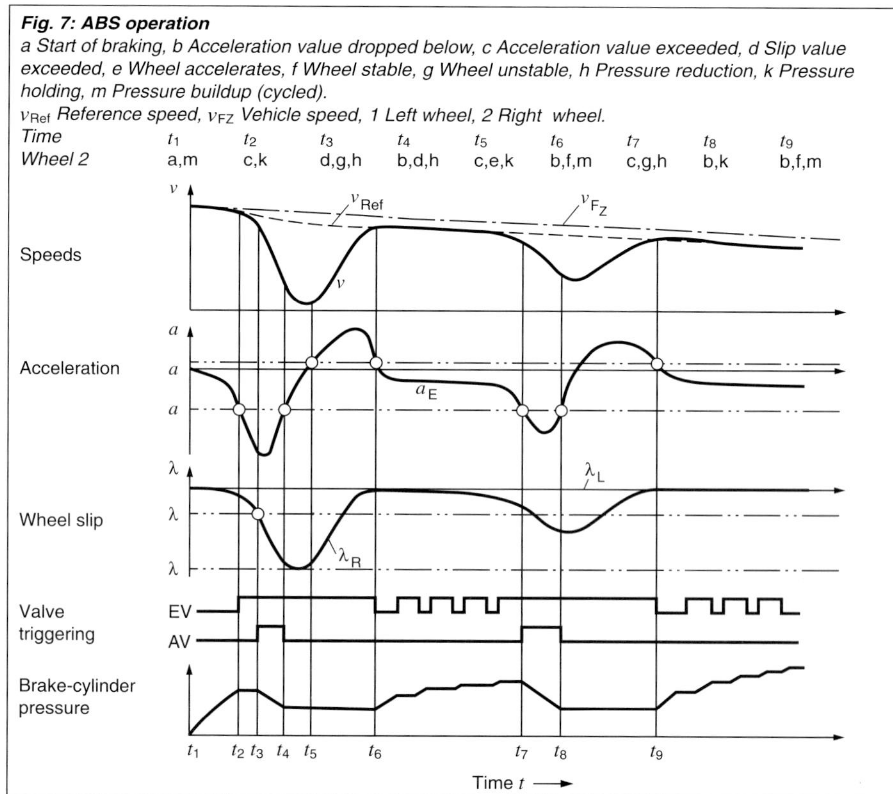

Fig. 7: ABS operation
a Start of braking, b Acceleration value dropped below, c Acceleration value exceeded, d Slip value exceeded, e Wheel accelerates, f Wheel stable, g Wheel unstable, h Pressure reduction, k Pressure holding, m Pressure buildup (cycled).
v_{Ref} *Reference speed,* v_{FZ} *Vehicle speed, 1 Left wheel, 2 Right wheel.*

Time	t_1	t_2	t_3	t_4	t_5	t_6	t_7	t_8	t_9
Wheel 2	a,m	c,k	d,g,h	b,d,h	c,e,k	b,f,m	c,g,h	b,k	b,f,m

tegy is based on a compromise in which the front wheel with the higher traction is braked somewhat less than possible, but still more than the front wheel with the lower surface adhesion. This modified control for the individual wheels reduces the amount of countersteer required to counteract the yaw tendency, with corresponding benefits regarding vehicle control. Brake pressure is controlled according to the basic schematic illustrated in Figure 7. The only difference being the limitation of brake pressure at the wheel with the higher road adhesion, and the resulting reduction in deceleration rate.

Applications

Commercial vehicles such as trucks and buses, tractor-semitrailer combinations, and conventional trucks with trailers, are all equipped with two, three or more axles. A variety of closed-loop control configuations are employed on such vehicles.

Figures 8 and 9 below illustrate the different types of control configuration. The code 4S/4K IRM is used as an example for explaining the designation codes:

– 4S indicates that wheel-speed sensors are installed at 4 wheels on this system.

– 4K means that the system incorporates four channels, with control of four pressure-control valves.

– IR: The system provides closed-loop control for each wheel (individual control). The system applies the optimum level of brake pressure to achieve the highest possible rate of deceleration (used on the rear axle).

– IRM: Modified version of individual closed-loop control. The extra feature (yawing-moment limitation) prevents the brake-pressure differential between left and right sides from exceeding a defined level (used on the front axle).

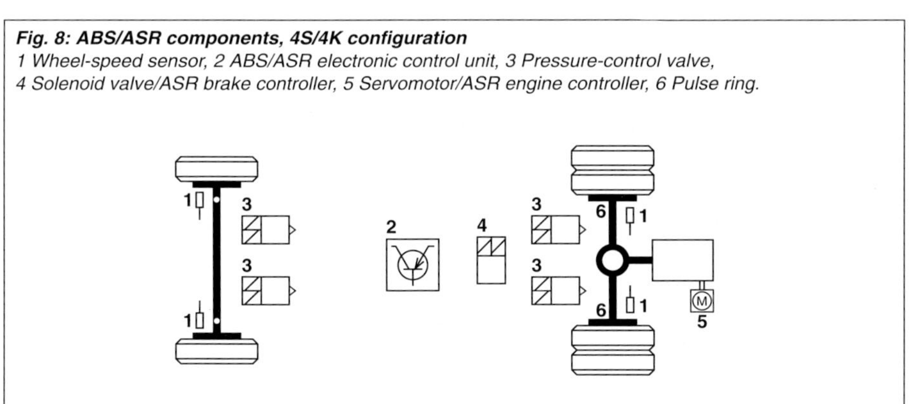

Fig. 8: ABS/ASR components, 4S/4K configuration
1 Wheel-speed sensor, 2 ABS/ASR electronic control unit, 3 Pressure-control valve,
4 Solenoid valve/ASR brake controller, 5 Servomotor/ASR engine controller, 6 Pulse ring.

Fig. 9: ABS/ASR components, 6S/6K configuration
1 Wheel-speed sensor, 2 ABS/ASR electronic control unit, 3 Pressure-control valve,
4 Solenoid valve/ASR brake controller, 5 Fuel-injection system, 6 Pulse ring, 7 EDC electronic control unit.

ASR traction control

Wheelspin can be encountered under a variety of conditions: During operation on slippery road surfaces affording only limited traction on one or both sides, while exiting ice-covered parking lots and road shoulders, during acceleration when cornering, or when accelerating from rest on steep slopes. When spinning – as when locked – the wheel provides only a limited amount of lateral adhesion (instability). Freely spinning wheels also cause wear in both tires and driveline components (such as the differential), with severe loads occurring when the wheel suddenly grabs when it hits a high-traction surface. ASR ensures optimal application of tractive forces by preventing wheelspin. ASR is an extension of ABS, with which it shares numerous components. The force which can be transferred when starting off or during acceleration is (as for braking) a function of the slip between tire and road surface. When braking, a wheel will lock within a few tenths of a second, while the build-up of excess torque during acceleration leads to a rapid increase in the rotational speeds of one or both drive wheels. ASR actually performs two functions: it enhances traction while maintaining vehicle stability (vehicle stays in lane)

Legal requirements

According to EU Directive 96/6/EEC, effective as of 1 January, 1994, trucks with an AGVW of 12 tons and above, and buses from 10 tons upward, must be equipped with a road-speed limitation facility. Maximum speed is 85 km/h for trucks and 100 km/h for buses. The ABS/ASR-M and ABS/ASR-P systems satisfy the requirements defined in the EU Directive.

Control concept

Brake control circuit
When the road surface provides different levels of traction on the left and right

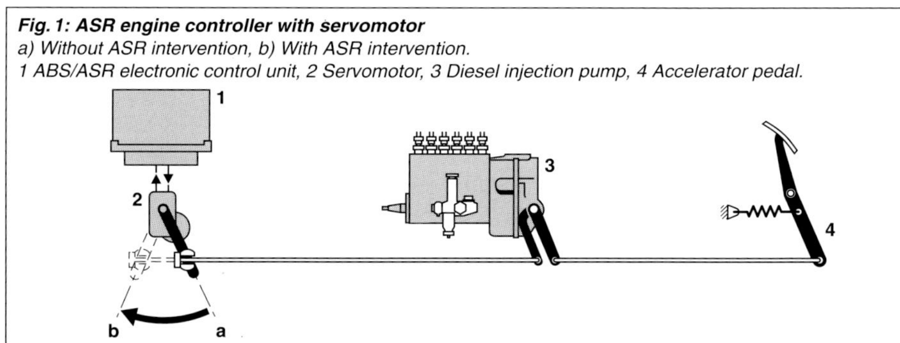

Fig. 1: ASR engine controller with servomotor
a) Without ASR intervention, b) With ASR intervention.
1 ABS/ASR electronic control unit, 2 Servomotor, 3 Diesel injection pump, 4 Accelerator pedal.

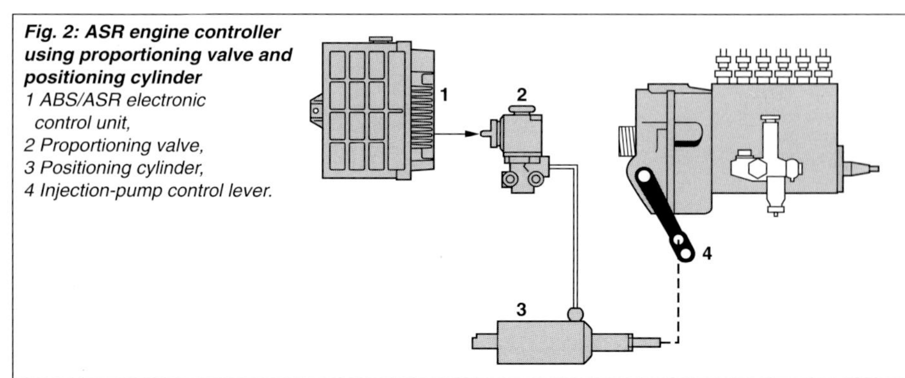

Fig. 2: ASR engine controller
using proportioning valve and
positioning cylinder
1 ABS/ASR electronic
*   control unit,*
2 Proportioning valve,
3 Positioning cylinder,
4 Injection-pump control lever.

sides, it will usually be only the wheel on the low-traction surface that spins. Under these circumstances it frequently proves impossible to transmit the required tractive force to the road, and the vehicle fails to pull away (Fig. 3).

The wheel-speed-sensor signals received by the ECU enable it to recognize impending wheel spin. It reacts by triggering the wheel's solenoid valve, so system pressure is directed to the pressure-control valve via the shuttle valve. The pressure-control valve relays modulated pressure to the brake cylinder, metering the pressure to allow the braking torque at the high-traction wheel to serve as drive (tractive) torque. In other words, ASR functions as an automatic limited-slip differential. During initial acceleration from a standing start, the brake controller operates up to a speed of 30 km/h. When the wheels lose traction at speeds in excess of 30 km/h, responsibility for corrective action reverts from the ASR brake-control circuit to the closed-loop engine-control circuit.

Closed-loop engine-control circuit

During standing start and acceleration, based on vehicle speed and the velocities of the drive wheels themselves, the ECU detects incipient loss of traction at the drive wheels and responds to this condition by reducing engine torque to the ideal level. The ECU can apply various strategies for engine intervention. Potential actuators, or final controlling elements, include:
- Electronic throttle control (ETC),
- Electronic diesel control (EDC),
- Proportioning valve with positioning cylinder P, Fig. 2,
- Servomotor M, Fig. 1, and
- Linear actuator.

When systems M and P are employed, the ECU directly reduces the drive torque at the diesel engine's fuel-injection pump by means of a servomotor or a valve with positioning cylinder. The other concepts rely on a reduction signal transmitted to the engine-management system. The ASR indicator lamp informs the driver of active ASR operation. This means that it functions as slip-warning indicator.

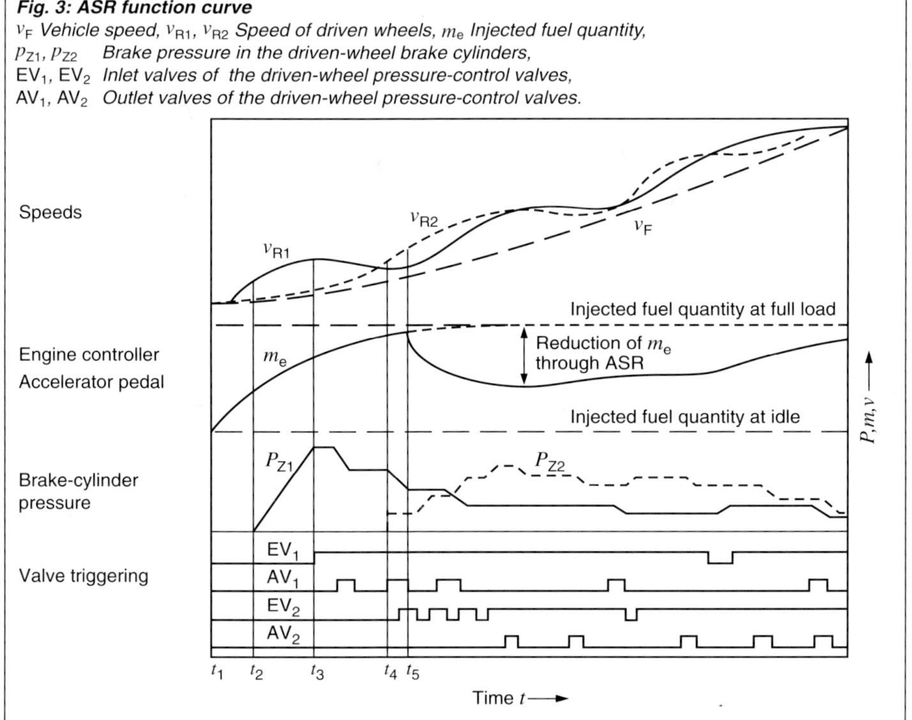

Fig. 3: ASR function curve
v_F Vehicle speed, v_{R1}, v_{R2} Speed of driven wheels, m_e Injected fuel quantity,
p_{Z1}, p_{Z2} Brake pressure in the driven-wheel brake cylinders,
EV_1, EV_2 Inlet valves of the driven-wheel pressure-control valves,
AV_1, AV_2 Outlet valves of the driven-wheel pressure-control valves.

Diagrams and descriptions of compressed-air systems

This section describes applications, design characteristics and operation of compressed-air systems. The descriptions are progressive, moving from basic to more complex structures, whereby the latter may incorporate features from those already dealt with.

Each system is based on the following component groups (Fig. 1):
– Component Group A
for compressed-air supply (generation and storage)
– Component Group B
service-brake system, ABS inclusive
– Component Group C
parking-brake system
– Component Group D
for trailer control
– Component Group E, pneumatic suspension
– Component Group F, door control
– Component Group G, traction control/ASR.

Component Groups A through C are the "standard equipment." They are the basic building blocks of every automotive compressed-air system. The other component assemblies are found only in more specialized vehicular applications, e.g., if the vehicle is equipped for trailer towing, includes pneumatic suspension or pneumatically-operated door control, or features ASR as a supplementary safety option.

The individual components in the six pneumatic systems illustrated on the following pages are represented by the graphic symbols defined in DIN 74 253. These units are dual-circuit and dual-line compressed-air power-brake systems, which operate with either a high or a low-pressure compressed-air supply.

The codes (DIN 74 254) used with the graphic symbols to identify connections can be identified using the first digit as a key:

0	Suction (air compressor)
1	Energy input
2	Energy output (except discharge to atmosphere, see Code 3)
3	Connection to atmosphere
4	Control connection (input at unit)
5	not assigned
6	not assigned
7	Antifreeze connection
8	Lubricant connection (air compressor)
9	Coolant connection (air compressor)

A second digit is added when several connections sharing a single function are present, e.g.:

21	Energy output to energy-storage device (compressed-air reservoir)
22	Energy output (switching connection).

If one connection is designed to perform several functions, it must be identified using two (initial) digits. These are separated by a hyphen. Example: 1-2 can serve as either an energy input or an energy output connection.

The symbols defined in DIN 74 253 are also used for lines and connections between the individual components:

— Pneumatic line, with no indication of flow direction
–▷ Pneumatic line, with indication of flow direction
–▶ Hydraulic line, with indication of flow direction
Electrical lead or line
═ Linkage and mechanical connections.

Not all of the brake systems described in the following operate exclusively on compressed air. General practice is reflected in portraying brake systems for light single vehicles (commercial vehicle without connections for trailer operation) that include hydraulic and mechanical control and transmission devices.

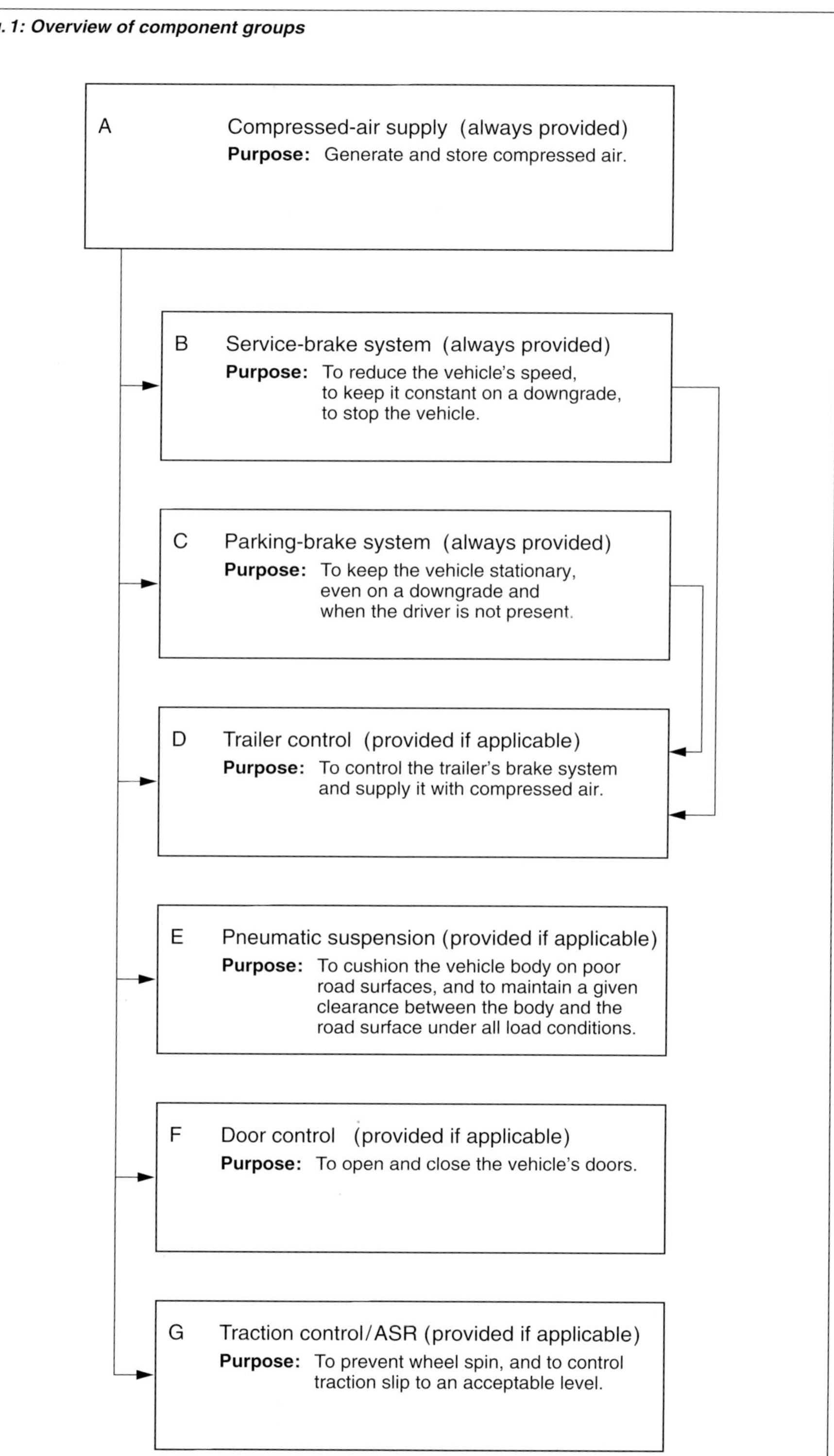

Fig. 1: Overview of component groups

A   Compressed-air supply (always provided)
Purpose: Generate and store compressed air.

B   Service-brake system (always provided)
Purpose: To reduce the vehicle's speed,
to keep it constant on a downgrade,
to stop the vehicle.

C   Parking-brake system (always provided)
Purpose: To keep the vehicle stationary,
even on a downgrade and
when the driver is not present.

D   Trailer control (provided if applicable)
Purpose: To control the trailer's brake system
and supply it with compressed air.

E   Pneumatic suspension (provided if applicable)
Purpose: To cushion the vehicle body on poor
road surfaces, and to maintain a given
clearance between the body and the
road surface under all load conditions.

F   Door control (provided if applicable)
Purpose: To open and close the vehicle's doors.

G   Traction control/ASR (provided if applicable)
Purpose: To prevent wheel spin, and to control
traction slip to an acceptable level.

System 1:

Dual-circuit compressed-air power-brake system

Low-pressure brake system featuring hydraulic transmission and mechanical parking brake. For trailerless commercial vehicles with an AGVW of between 6 and 9 metric tons.

Design

System 1 is composed of the "basic equipment" included in the first three component groups (A, B, C, see Fig. 1):

– The compressed-air supply (A) compresses air (8 bar) for storage in two pneumatic supply circuits. This air is then supplied to the service-brake system.

– The service-brake system (B) controls the compressed air during braking by means of its downstream hydraulic circuit, in order that the brakes are applied at all of the vehicle's wheels.

– The parking-brake system (C) operates the brakes at the rear axle via linkage rods and Bowden cable.

Operation

Compressed-air supply (A)

After being drawn-in and compressed in the compressor (1), the air passes through the low-pressure (set to approx. 8 bar) pressure regulator (2) used to control air pressure levels in the air tanks/reservoirs (4) before arriving at the dual-circuit protection valve (3).

The dual-circuit protection valve distributes the compressed air emerging from the pressure regulator to the two compressed-air supply circuits (21 and 22). The safety valve also ensures that the two circuits remain mutually isolated; failure in one of the supply circuits will not impair operation of the undamaged circuit.

Upon leaving the two-circuit protection valve, the air flows to the two air tanks/reservoirs (4) for storage. Should the system pressure drop below the mini-mum approved level, the pressure switch (6) closes to actuate an acoustic and optical warning signal. Either manual or automatic drain valves (5) are employed to drain condensation from the air tanks.

Service-brake system (B)

The service brakes (B) function as a compressed-air power-brake system featuring a hydraulic transmission (force-transfer) device. The driver applies the service brakes at both front and rear axles by pressing on the brake pedal.

The amount of force applied to the pedal determines the rate at which compressed air flows from the air tanks (4) and into the dual-circuit brake assembly (7) – a combination dual-circuit service-brake valve and dual-circuit actuator cylinder for the brake master cylinder. The force that the air exerts against the cylinder's working piston is proportional to that applied at the brake pedal. This mechanical force is converted to hydraulic pressure in the flange-mounted tandem master cylinder (8). Finally, the hydraulic pressure is conveyed through the brake circuits to the wheel-brake cylinders (9) at the front and rear axles, extending the brake shoes and pressing them against the brake drums.

Secondary brake system
The service-brake system also assumes the secondary (or emergency) braking function in the event of failure in one of the air-supply or hydraulic circuits. Failure in one of the air-supply circuits will not affect the braking force transferred to the wheel-brake cylinders (9). If one of the hydraulic brake circuits fails, braking remains available on the vehicle axle connected to the intact circuit. The system remains capable of supplying the minimum braking required from a secondary brake system.

Parking-brake system (C)

The underlying principle of the parking-brake system (C) is operation based on muscular energy. The driver operates this system using a hand-brake lever (10) to apply the rear wheel brakes via mechanical linkage.

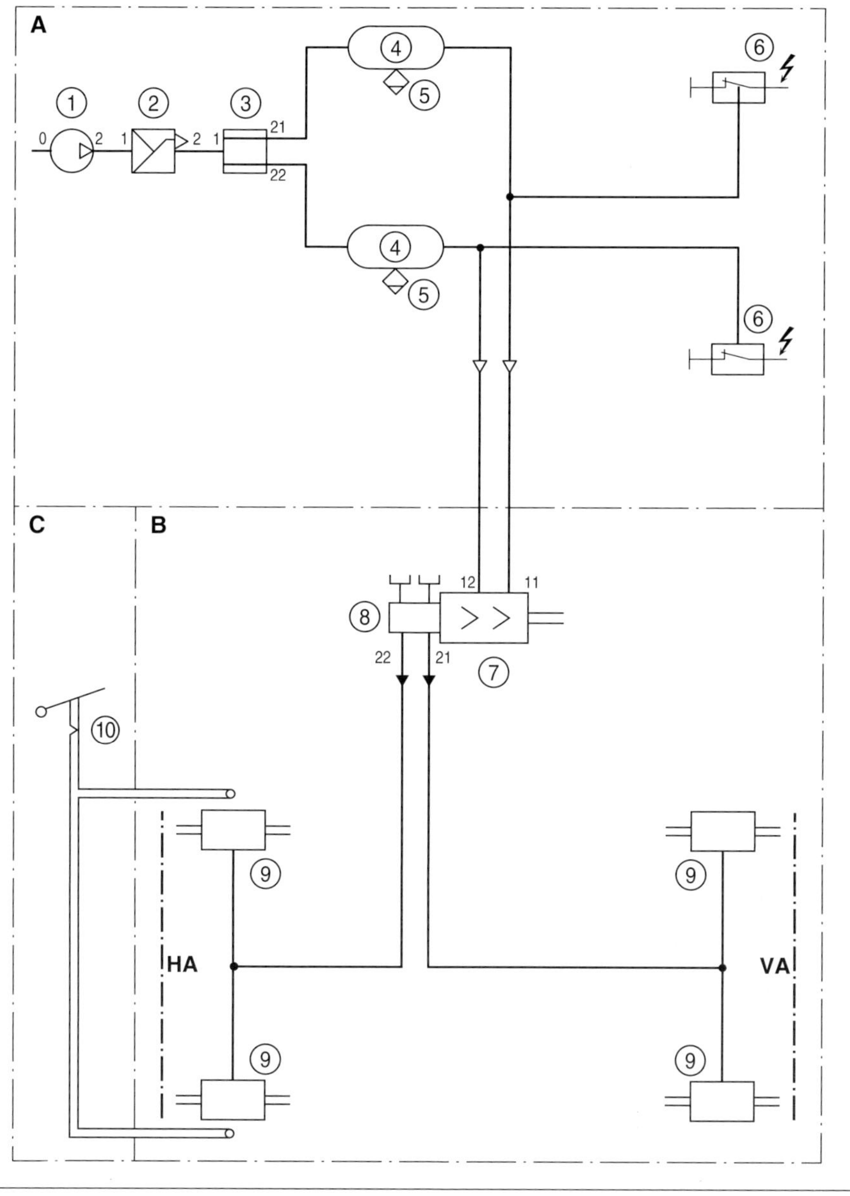

Fig. 1: Equipment groups and individual units
A Compressed-air supply
1 Air compressor, 2 Pressure regulator, 3 Dual-circuit safety valve, 4 Air reservoir, 5 Drain valve, 6 Pressure switch (nc contact).

B Service-brake system
7 Dual-circuit brake assembly, 8 Tandem master cylinder (hydraulic), 9 Wheel-brake cylinder (hydraulic).

C Parking-brake system
10 Hand-brake lever.

For reasons of clarity, the right front axle is shown. VA Front axle, HA Rear axle.

System 2:

Dual-circuit compressed-air power-brake system

High-pressure brake system featuring spring-type brake actuators for the parking brake.
For trailerless commercial vehicles with an AGVW of between 6 and 13 metric tons.

Design

This system (Fig. 1) shares its basic components with System 1:

– Compressed-air supply (A),
– Service-brake system (B), and
– Parking-brake system (C).

At the same time, it differs from System 1 in the following respects:
– Four compressed-air supply circuits with two air tanks (instead of two compressed-air supply circuits with two air tanks),
– One four-circuit protection valve (instead of a two-circuit protection valve),
– A supplementary antifreeze pump,
– A additional hydraulic load-sensing valve (braking-force regulator), and
– A compressed-air parking-brake system of the power-brake type, featuring non-mechanical (no linkage) operation (instead of a mechanical parking brake operated by muscular force).

Operation

Compressed-air supply (A)

First, the air is drawn in and compressed by the air compressor (1). It then flows through the high-pressure (set to 16 bar) pressure regulator (2), which controls the air pressure in the two compressed-air tanks/reservoirs (5). Form there, the air continues to the automatic antifreeze pump (3), which meters antifreeze to the compressed air as required. During winter operation, every time the system is switched on, the pressure regulator (2) directs a pressure pulse to the automatic antifreeze pump, which responds by injecting antifreeze to prevent condensate from freezing within the system.

After leaving the antifreeze pump the compressed air flows to the four-circuit protection valve (4). This valve controls air distribution while at the same time isolating the individual circuits from each other: These comprise the compressed-air supply circuits 21 and 22 for the service-brake system, supply circuit 23 for ancillary equipment a (such as pneumatic control for the engine brake) and supply circuit 24 for the parking- brake system.

The pressure switches (7) respond to low system pressure by triggering an acoustic or optical warning signal.

The drain valves (6) are used to release condensation that collects in the compressed-air tanks.

The high-pressure operating concept makes it possible to use smaller air tanks. Because the ancillary equipment operates on low pressure, a pressure limiter (8) serves as a reduction unit, converting the air to low pressure (8 bar).

Service-brake system (B)

The service brakes operate according to the same basic principles as described in the previous brake system (System 1); this is also a compressed-air power-brake system with a hydraulic transmission (force-transfer) device.

This system responds to driver-applied pressure at the brake pedal, and operates the brakes on both front and rear axles.

The force applied at the brake pedal determines the quantity of compressed air (and the brake pressure) that flows from the air tanks (5) assigned to the brake system. This air flows through the service-brake valve (9) and into the dual-circuit actuating cylinder (10), where it acts upon the hydraulic tandem master cylinder (11). Here the mechanical force applied by the actuating cylinder (10) is converted into hydraulic pressure for transmission through the respective brake circuits and to the wheel-brake cylinders (12) at the front and rear axles.

Fig. 1: Equipment groups and individual units
A Compressed-air supply (high-pressure system)
1 Air compressor, 2 Pressure regulator, 3 Antifreeze pump, 4 Four-circuit protection valve, 5 Air reservoir, 6 Drain valve, 7 Pressure switch, 8 Pressure limiter.

B Service-brake system
9 Service-brake valve, 10 Dual-circuit actuating cylinder for brake master cylinder, 11 Tandem master cylinder (hydraulic), 12 Wheel-brake cylinder (hydraulic), 13 Load-sensing valve (controlled from the air suspension).

C Parking-brake system
14 Parking-brake valve (with pressure limiter), 15 Relay valve, 16 Spring-type brake actuator, a secondary loads.

VA *Front axle,* HA *Rear axle.*

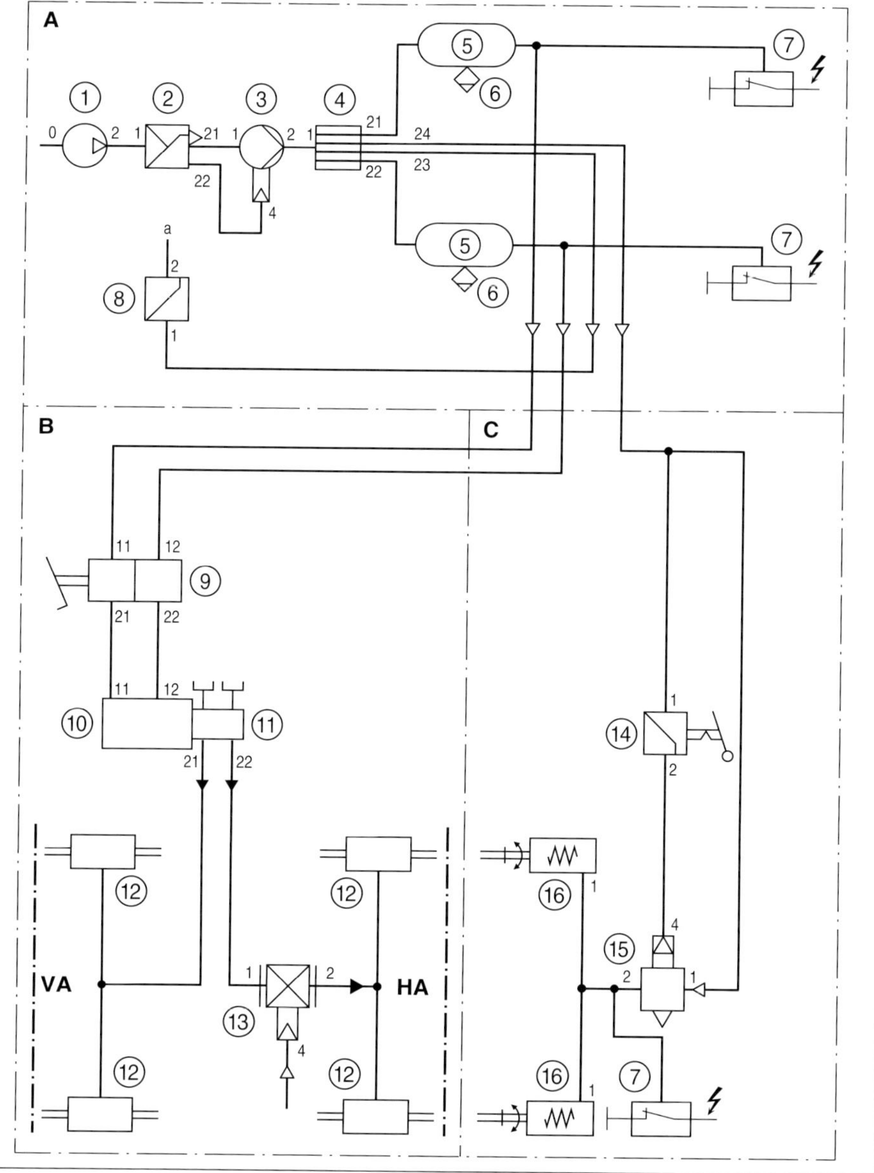

System 5:

Dual-line compressed-air power-brake system

With ABS.
Low-pressure brake system with spring-type brake actuator for the parking brakes.
For commercial vehicles with trailer operation and an AGVW in excess of 16 metric tons.

Design

The system is composed of four component groups (Fig. 1, Page 113):
– Compressed-air supply
(Component Group A) generates compressed air at standard pressure (8 bar). It then stores this compressed air for supply to:
– The service-brake system
(Component Group B),
– The parking-brake system
(Component Group C), as well as
– To the trailer brakes through the trailer-control circuit (Component Group D).

The differences relative to System 4 are as follows:
– Closed-loop control of braking force also operates on front wheels,
– 3-axle 6 x 2 commercial vehicles, and
– Incorporates ABS.

Operation

Compressed-air supply (A)
The system employed to deliver pneumatic pressure is basically the same as that in System 3. The four-circuit protection valve (4) distributes the compressed air among four individual circuits:
– Circuits 21 and 22 for the service-brake system,
– Circuit 23 for the parking-brake system and trailer supply, and
– Circuit 24 for ancillary equipment a.

Service-brake system (B)
When the driver actuates the service-brake valve (8), compressed air flows from the air tanks/reservoirs (5) and through connection 22 on its way to the front axle. The compressed air passes through the open pressure-control valves (14), proceeding unhindered to the diaphragm actuators (9).
The service-brake valve (8) controls, via connection 21, the load-sensing valve (10) which conducts the air directly from the air tank/reservoir (5) to the diaphragm actuators (9) on the leading and trailing axles as well as to the diaphragm actuators on the drive axle.
The service-brake valve governs the trailer-control valve (18) through connections 21 and 22. The mechanical load-sensing valve (10) automatically adapts the brake pressure at the drive and leading axles to correspond to axle loads while simultaneously using the proportioning relay in the service-brake valve (8) to reduce brake pressure at the front axle.
When braking, if the signals from the wheel-speed sensors (13) indicate to the ECU (12) that one or more wheels are on the verge of locking; the ECU responds by activating the pressure-control valves (14) for the affected wheel(s).
This reduces brake pressure at the potentially locking wheel(s) so that wheel lock is prevented. The drive axle also controls the leading axle.

Parking-brake system (C)
Design and function correspond to those of the assembly used in System 4 above.

Trailer control (D)
Supply and control layout for the trailer brakes correspond to those described for System 4 above.

Fig. 1: Equipment groups and individual units
A Compressed-air supply
1 Air compressor, 2 Pressure regulator, 3 Single-box air drier with 3a regeneration-air tank, 4 Four-circuit protection valve, 5 Air reservoir, 6 Drain valve, 7 Pressure switch, a Secondary loads.
B Service-brake system
8 Service-brake valve (with adapter valve), 9 Diaphragm actuator, 10 Load-sensing valve (with relay valve), 11 Combination brake cylinder, 12 Electronic control unit (ABS), 13 Wheel-speed sensor (ABS), 14 Pressure-control valve (ABS).
C Parking-brake system
15 Non-return valve, 16 Parking-brake valve, 17 Relay valve.
D Trailer control
18 Trailer-control valve, 19 Coupling head "Supply", 20 Coupling head "Brakes"

For reasons of clarity, the front axle is shown on the right. VA *Front axle*, HA *Rear axle*,
3A = 3rd axle = Lifting axle.

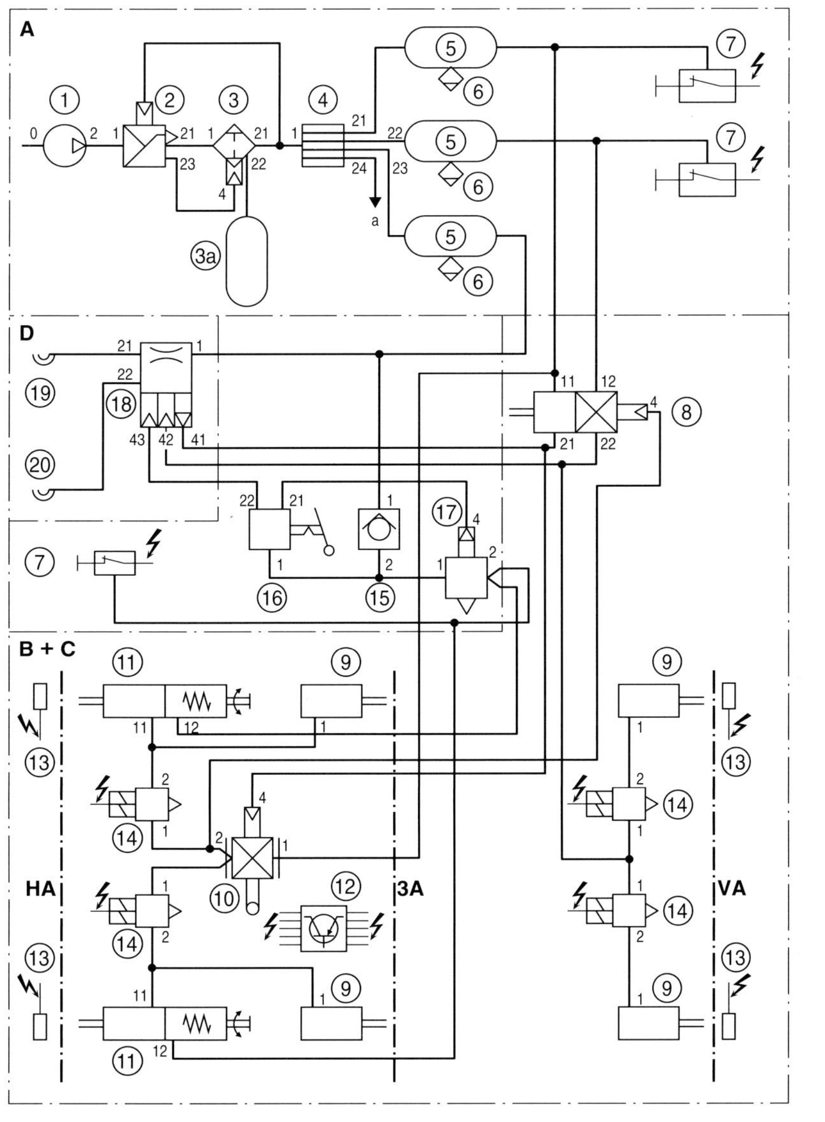

Equipment for commercial vehicles

Component groups

Every compressed-air brake system is constructed in modular form using component groups:

- Component group A:
  Compressed-air supply (compressed-air generation and storage)
- Component group B:
  Service-brake system
- Component group C:
  Parking-brake system
- Component group D:
  Trailer control
- Component group E:
  Pneumatic suspension
- Component group F:
  Door control
- Component group G:
  Traction control (ASR).

The component groups A to C are the "standard equipment" and are a part of all compressed-air brake systems. In contrast, the other component groups are only installed if the vehicle is equipped for operation with a trailer, if it is provided with pneumatic suspension, if the doors are operated pneumatically, or if traction control (ASR) is fitted as a safety accessory.

The individual components are described in the sequence specified by the component groups A to G. Normally, they are illustrated by sectional views (in part with different operating positions).

The first digit of the connection designations according to DIN 74 254 used in the graphic symbols has the following meaning:

0	Intake port (air compressor)
1	Energy supply
2	Energy outlet (not for outlet to the atmosphere; see number 3)
3	Port to the atmosphere
4	Control port (the input at the component)
5	Unassigned
6	Unassigned
7	Antifreeze port
8	Lubricating-oil port (Air compressor)
9	Cooling-water port (Air compressor)

A second digit is provided if there are a number of ports of the same type, e.g.

21	Energy outlet to the energy storage device (compressed-air reservoir)
22	Energy outlet (switching port)

If one port can perform several functions, it must be designated using two (initial) digits. These are to be separated from one another by means of a dash, e.g. 1-2, optionally the energy supply or the energy outlet.

For the leads, lines and connections between the individual components, the graphic symbols of DIN 74 253 are likewise used.

—— Pneumatic line without a specified direction of flow
—▷ Pneumatic line with a specified direction of flow
—► Hydraulic line with a specified direction of flow
⌁ Electric lead or cable
══ Linkage and mechanical connection

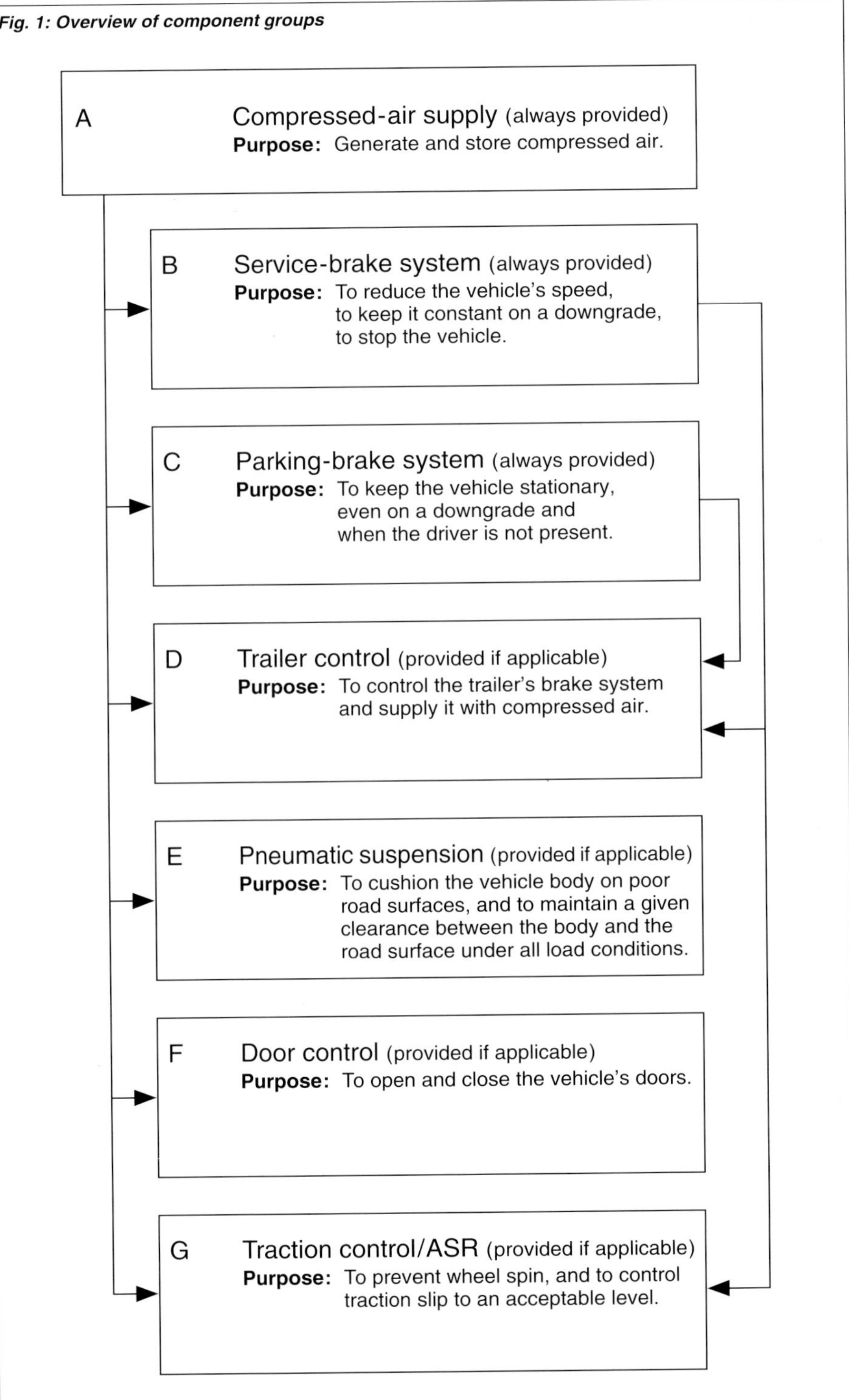

Fig. 1: Overview of component groups

A Compressed-air supply (always provided)
Purpose: Generate and store compressed air.

B Service-brake system (always provided)
Purpose: To reduce the vehicle's speed,
to keep it constant on a downgrade,
to stop the vehicle.

C Parking-brake system (always provided)
Purpose: To keep the vehicle stationary,
even on a downgrade and
when the driver is not present.

D Trailer control (provided if applicable)
Purpose: To control the trailer's brake system
and supply it with compressed air.

E Pneumatic suspension (provided if applicable)
Purpose: To cushion the vehicle body on poor
road surfaces, and to maintain a given
clearance between the body and the
road surface under all load conditions.

F Door control (provided if applicable)
Purpose: To open and close the vehicle's doors.

G Traction control/ASR (provided if applicable)
Purpose: To prevent wheel spin, and to control
traction slip to an acceptable level.

Component group A "Compressed-air supply"

Purpose

To generate and store compressed air (Fig. 1).

Air compressor

Purpose
The air compressor (2) is the source of energy of a compressed-air brake system: The required compressed air is generated here.

Design
An air compressor is a piston pump (Fig. 3). The crankshaft (7) of this pump is driven by the vehicle engine through a V-belt or a toothed gear. An air compressor essentially consists of the crankcase (6) with the crankshaft and bearing as the drive element, the cylinder (3) with piston (4) and connecting rod (5) for the intake and compression of the air, the intermediate plate (2) with flap valves for the inlet and outlet and the cylinder head (1) with the air ports for intake air and compressed air (in the case of water-cooled compressor cylinder heads, water ports are also present).

In view of the fact that only filtered air may be used, the intake line is connected to the engine air filter.

The air compressor is provided with either an air-cooled or a water-cooled cylinder head. With the latter version, in order to improve cooling, coolant also flows through the intermediate plate.

The air compressor is connected to the engine lubrication circuit.

Fig. 2: Air compressor

Fig. 1: Component group A "Compressed-air supply"

A	Compressed-air supply					

B	C	D	E	F	G
Service-brake system	Parking-brake system	Trailer control	Pneumatic suspension	Door control	Traction control (ASR)

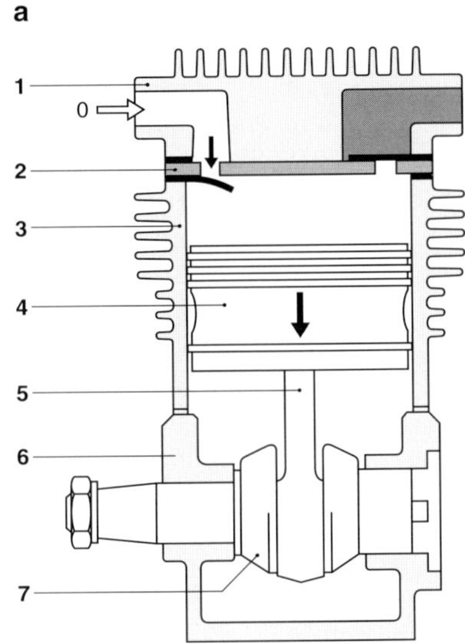

Fig. 3: Air compressor, air-cooled (sectional view)
a) Intake,
b) Compression and feed.
1 Cylinder head,
2 Intermediate plate (with inlet and outlet valves),
3 Cylinder,
4 Piston,
5 Connecting rod,
6 Crankcase,
7 Crankshaft.

a

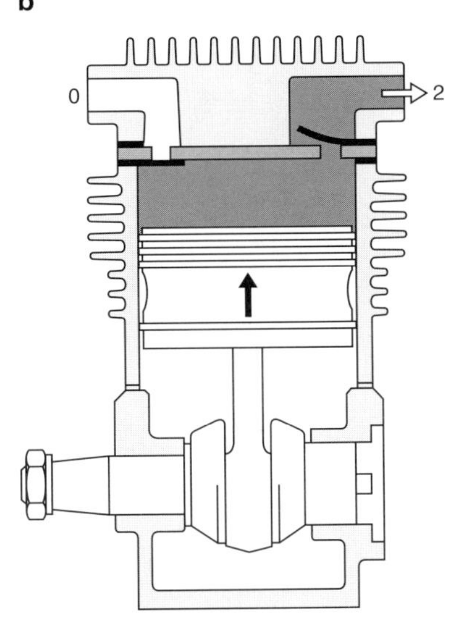

b

Operating principle

As the piston retracts, the inlet valve opens, allowing air to enter. During the upward stroke, the piston compresses this air. As soon as a certain pressure has been reached, the outlet valve opens and the compressed air is fed to the pressure regulator.

The air compressor runs as long as the vehicle engine is running. Once the compressed-air brake system has been filled, the air compressor feeds the air in an almost pressureless state to the atmosphere by way of the pressure regulator which is switched to the idle position.

Pressure regulator

Purpose

The pressure regulator conducts the compressed air, which is constantly supplied by the air compressor as long as the vehicle engine is running, either to the air reservoirs or to atmosphere. In this manner, it regulates the supply pressure of the compressed-air brake system which remains within the operating range, i.e. between the cut-in pressure and the cut-off pressure of the pressure regulator.

The tire-inflation fitting installed in the pressure regulator is used for inflating vehicle tires and for filling the compressed-air brake system from an external source.

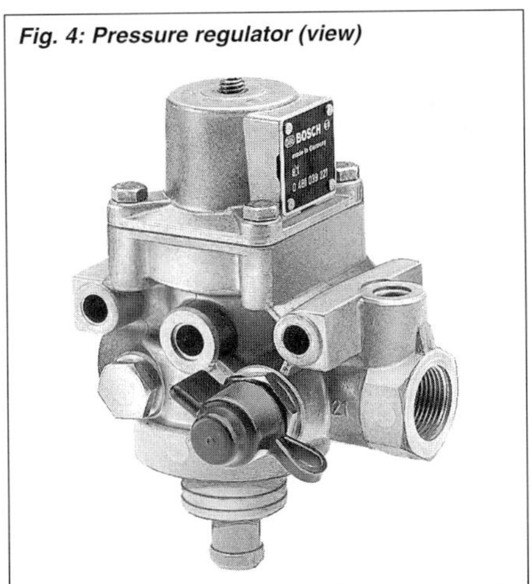

Fig. 4: Pressure regulator (view)

Design

The pressure regulator (Fig. 4) is a combination of a pressure regulator, air filter, safety valve, check valve and tire-inflation device unified in a single housing. The schematic sectional view (Fig. 5) depicts the most important parts for the regulation of pressure:

At the top, the control valve (5), actuated by the diaphragm (2), at the bottom the idle valve (7) which is actuated by the piston (6) and, at the bottom right, the check valve (15). The check valve is omitted on pressure regulators which control a downstream air dryer.

In addition, the tire-inflation fitting 1-2 with double-seat valve (10,11) can be seen. The tire-inflation fitting is not provided on all pressure regulators. In addition to the ports 1 for air supply and 21 for the outlet of air to the air reservoir or air dryer, some pressure regulators have a port 22 which can be used to tap the supply pressure of the air compressor, for example, for the pulse control of the antifreeze pump.

Port 23 (not shown) controls the air dryer's idle piston while port 4 (not shown) measures the pressure downstream of the air dryer.

The loud noise of escaping air which usually results when the pressure regulator switches from "Filling" to "Idle" can be reduced by means of an elongated blow-off fitting (8) fitted with a silencer.

Operating principle

Filling

The compressed air entering at port 1 passes through the channels (12) and (13) upstream of the check valve (15), opens this valve and proceeds via port 21 through the pressure line to the air reservoir until the cut-off pressure has been reached.

Idle

The cut-off pressure exceeds the force of the helical compression spring (1) and thus forces the diaphragm (2) upward. In doing so, the cup seal of the control valve (5) is lifted off of the control valve's seat by means of the hollow pin (3).

Fig. 5: Pressure regulator (sectional view)
a) Filling mode, b) Idle mode.
1 Compression spring, 2 Diaphragm, 3 Pin, 4 Compression spring, 5 Cup seal (control valve),
6 Piston, 7 Valve plate (idle valve), 8 Blow-off fitting, 9 Bleed, 10 Tappet, 11 Valve cone, 12 Air passage,
13 Air passage, 14 Air passage, 15 Valve cone (check valve), 16 Compression spring.

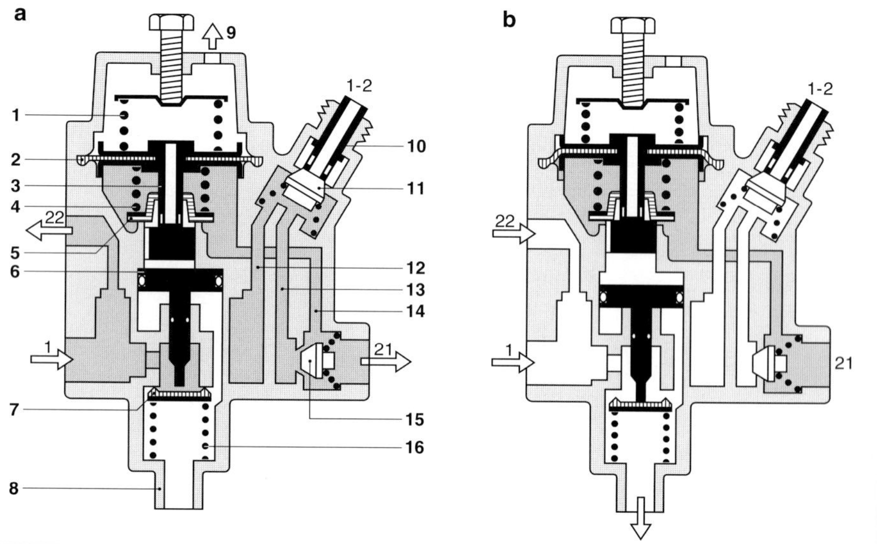

The compressed air now flows through the control valve (5) and presses the piston (6) against the force of the compression spring (16) so that the idle valve (7) opens. The air which is constantly being supplied by the air compressor now flows to atmosphere via the idle valve (7) while the check valve (15) remains closed. This prevents air from flowing back out of the reservoirs. If the pressure in the brake system drops, e.g. by actuating the brake equipment, until the cut-in pressure is reached, the force of the helical compression spring (1) acting from above on the diaphragm (2) is greater than the force of the compressed air acting from below. As a result, the diaphragm (2) moves down again and the spring (4) pressing on the cup seal closes the control valve (5).

The compressed air above the piston (6) escapes to atmosphere by way of the hollow pin (3), the compression-spring chamber, and the bleed (9) below the rating plate.

The piston (6) is pushed upward by the force of the spring (16) and the idle valve closes. The air compressor now supplies air to the air reservoir again. The pressure of the air supplied by the air compressor remains between the cut-in pressure and the cut-off pressure. The difference between these two pressures is known as the operating range. When the cut-off pressure is reached, the idle valve opens immediately.

Tire inflation

It is only possible to inflate tires with the pressure regulator in the filling position. The pressure regulator can be made to switch from the idle position to the filling position by pressing the brake pedal.

When the tire-inflation hose is screwed into the tire-inflation fitting (Fig. 6), the hollow tappet (10) is pushed back. In so doing, it presses the valve cone (11) up against the valve seat on the other side.

Due to this, no more compressed air can reach the air reservoirs. The compressed air now flows into the tire-inflation hose. In view of the fact that the control valve

(5) is bypassed during tire inflation, the idle valve (7) opens when the relief pressure has been reached.

Filling the brake system from an external compressed air system

When the engine is stopped or on an assembly line, the brake system can also be filled with compressed air by way of the tire-inflation fitting. However, the tire-inflation fitting 1-2 must not be screwed in so far that the tappet (10) presses the valve cone (11) up against the valve seat on the other side. There would then be no connection between the pressure-regulator inlet (port 1), the outlet (port 21), and the tire-inflation fitting (ports 1-2).

Protection against excessive pressure

If the pressure regulator is defective or frozen, or when tires are being inflated, pressure can rise to an impermissibly high level. In this case, the idle valve (7) acts as a safety valve when the relief pressure has been reached and allows the compressed air to escape to atmosphere.

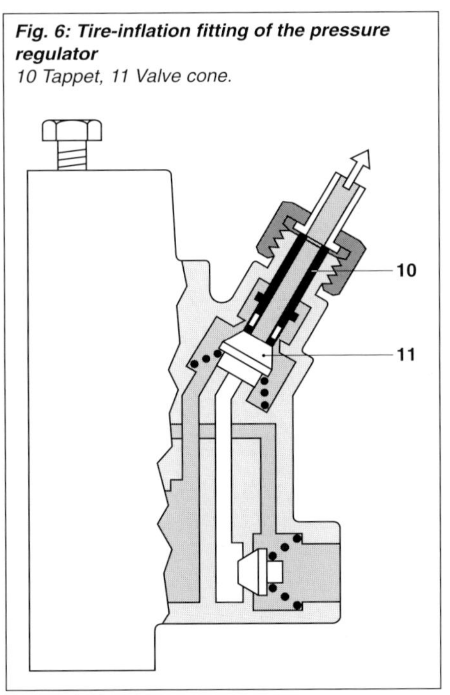

Fig. 6: Tire-inflation fitting of the pressure regulator
10 Tappet, 11 Valve cone.

Air dryers

Why are they necessary?
Compressed air represents an optimal energy source and an ideal force-transmission medium for brake systems on medium-heavy and heavy-duty commercial vehicles. However, one disadvantage is the tendency of air to absorb moisture in the form of water vapor. This water vapor condenses once the air, which is heated during compression, cools back to ambient temperature upon emerging from the air compressor. The resulting condensation assumes the form of moisture in pneumatic lines, air tanks, cylinders and valve assemblies, and can have negative consequences for the brake system and vehicle safety:
– In winter, this water can freeze, leading to the risk of malfunctions in valve assemblies,
– it can promote corrosion in air tanks, cylinders and valve assemblies, and
– it can destroy the protective lubricant film that is applied to brake-system components.

Originally, efforts to combat condensation in pneumatic systems were largely restricted to passive countermeasures:
– manually-operated or automatic drain valves to discharge condensate from air cleaners and tanks, and
– antifreeze pumps to inject antifreeze, in order to lower the condensate freezing point.
Systems designed to dry the compressed air, preventing condensation from the outset, are far superior to the passive measures listed above.

Drying principle
The pneumatic systems on newer-model vehicles are equipped with air dryers, in which the air supply is dried according to the adsorption principle. In these systems, the compressed air flows through a granular dessicant with molecular sieves featuring a crystalline structure which captures the water molecules (adsorption, Figure 7). The dessicant's moisture-retention capacity increases along with falling air temperature and increasing air pressure (Figure 8).
The fact that the dessicant absorbs far less water at atmospheric pressure is exploited for the regeneration process.
Here, a portion of the dried air is discharged through a regeneration throttle and conducted through the moist granulate in a reverse-flow current. The dry air then absorbs a large proportion of the moisture trapped in the crystalline structure as it flows through the molecular sieve.
At the beginning of the drying process, due to the expansion which occurs when the pressure regulator changes from

Fig. 7: Dessicant (granulate)
Schematic of the molecular sieve.
1 Air, 2 Surface, 3 Macropores, 4 Micropores, 5 Water molecule.

Fig. 8: Water-absorption capacity of the dessicant as a function of pressure and temperature

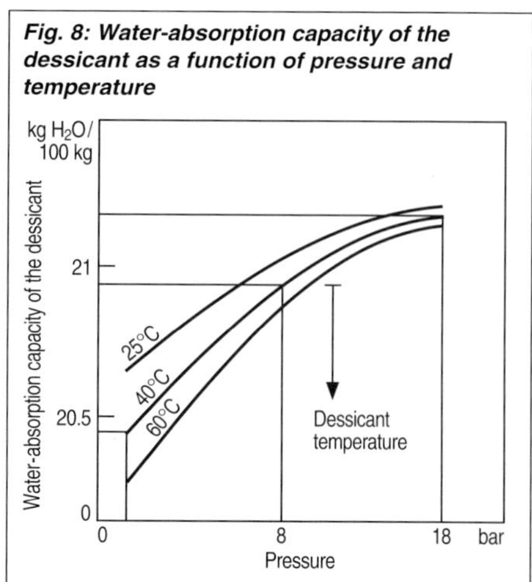

"charge" to "idle", a substantial portion of the water contained on and adjacent to the surface of the molecular sieve's macropores, in the form of supersaturated water vapor, is captured and transported to the outside atmosphere.

Lowering the dew point

At a given pressure, the dew point is defined as the temperature at which the air's relative humidity is 100 %, corresponding to maximum saturation.

In cold, temperate and warm tropical climate zones the temperature can fall up to 20 °C (36 °F) within a period of twelve hours.

Thus, to maintain the dew point at a sufficiently low level to satisfy the requirements of the compressed-air circuit, the air dryer must be capable of removing enough moisture from the air to prevent condensation from forming in response to temperature dips of this magnitude.

The air dryer and its ancillaries should be designed with a safety margin of 5...10 °C (lowering the dew point by a cumulative 25...30 °C), as corrosion can occur once the relative humidity exceeds 60 %.

Water content of air

The given conditions define the specific level of (gaseous) water vapor that air can absorb. Water vapor starts to condense once this maximum level is exceeded. The relative humidity index indicates the proportional relationship between the actual water-vapor concentration and the current saturation point (water content). The saturation point is defined by two parameters: pressure and temperature. When considering an air dryer's drying capacity, it is important to include the moisture content of the air being drawn into the air compressor in the calculations.

The intake air's moisture content responds to climatic variations by fluctuating throughout a large range. These fluctuations have an effect on the actual drop in dew point.

Evaluations based on actual measurements of atmospheric humidity in various European climate zones indicate an annual mean moisture content of 10 g H_2O per m^3 of air. Air can absorb this quantity of moisture at 8 bar and at the recommended maximum air-dryer intake temperature of 65 °C.

Once the air's moisture content exceeds this level, a portion of the water vapor condenses in the lines leading to the air dryer, while condensation also forms on the inside walls of the dessicant cartridge before the air can flow through the dessicant itself. When the pressure regulator cycles from "charge" to "idle" the compressed air expands and is conducted to the outside atmosphere.

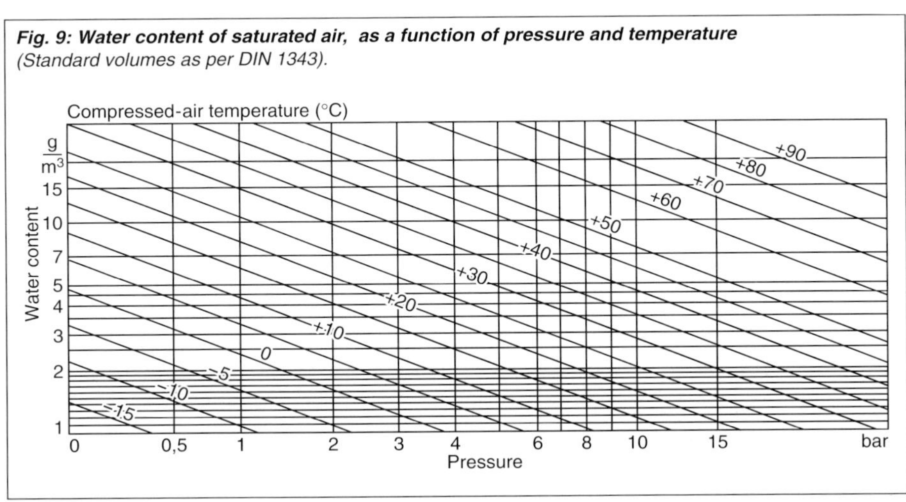

Fig. 9: Water content of saturated air, as a function of pressure and temperature
(Standard volumes as per DIN 1343).

Air-dryer design

The air dryer consists of the control section (assembly comprising components 1...6 in Figure 10) and the dessicant cartridge.

During the idle phase, the regeneration current flows through the bleeder valve (2) – which also serves as a safety valve – and the drain fitting (3) on its way to the outside atmosphere. The temperature at the bleeder valve is maintained by the heating element (1). A portion of the compressed air, specially stored for dessicant regeneration, is released by the throttle (4) and expands back to atmospheric pressure. The check valve (5) prevents reverse flow of compressed air from the downstream brake circuit.

The dessicant cartridge (6) contains a preliminary filter (9) designed to trap oil and air-borne particulates, a granulated dessicant (8) to remove the moisture from the compressed air as it passes, as well as a secondary filter (10) to retain dessicant particles, and a compression spring (7) to maintain the dessicant under pressure.

Operation

Drying process

Upon emerging from the pressure regulator, the air from the compressor flows through connection 1 and into the control stage (components 1...5) in the air dryer before proceeding into the dessicant cartridge (6), where it flows through the preliminary filter (9) and the dessicant (8). The preliminary filter retains coarse contaminants. It also assumes an initial drying function, based on the fact that the compressed air cools during its passage through this unit, forming condensation in response to its reduction in moisture-retention potential that accompanies the drop in dew point.

The primary drying process occurs as the air flows through the dessicant (8). Here, the high pressure levels promote a process designed to ensure that the compressed air deposits enough moisture on the dessicant to effectively prevent additional condensation from forming in the system, even under extreme climatic conditions.

Fig. 10: Air dryer without pressure regulator
a) Drying, b) Regeneration.
1 Heating element, 2 Bleeder valve, 3 Drain fitting, 4 Throttle, 5 Check valve, 6 Dessicant box,
7 Compression spring, 8 Dessicant cartridge, 9 Preliminary filter, 10 Afterfilter.
Connections: 1 From air compressor, 21 To air reservoir, 22 To regeneration-air reservoir, 3 Bleed,
4 From pressure regulator.

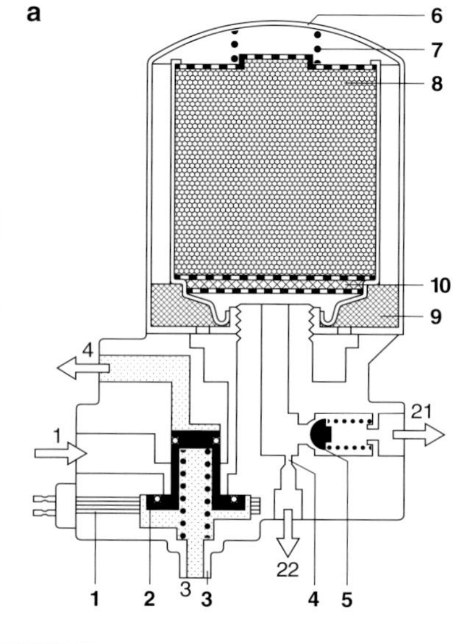

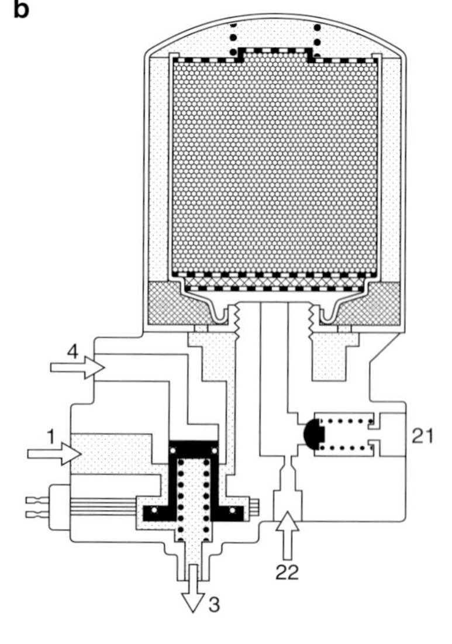

After drying, the compressed air flows through the secondary filter (10) and the check valve (5) before exiting through connection 21 on its way to the air tanks, or continuing through the throttle (4) and connection 22 on its way to the regeneration reservoir.

Regeneration

Whenever the pressure regulator returns to its idle state, the dessicant is regenerated; in this process a substantial proportion of the moisture captured in the dessicant is removed again.

Once the pressure in the brake circuit reaches the corresponding cut-off point, the pressure regulator reverts to the idle state while at the same time applying pressure at connection 4 on the air dryer to open the bleeder valve (2). This releases the pressure on the compressed air in the dessicant box (6): compressed air and condensate escape to atmosphere through the drain fitting (3).

At the same time, dehumidified compressed air from the regeneration reservoir flows through connection 22 to the throttle (4), where its moisture-absorption potential increases as it expands back to atmospheric pressure.

Upon emerging from the secondary filter (10), the air current flows through the saturated dessicant, where it absorbs a substantial proportion of the captured moisture before carrying the evaporate through the drain fitting (3) on its way to the outside atmosphere. This process of moisture transfer regenerates the dessicant, preparing it to remove moisture from the next charge of humid compressed air when the air dryer switches over to its drying mode again.

The air that flows through the air dryer's bleeder valve (2) contains water. A heating element (1), designed for automatic activation once the temperature falls below a specified point, is installed to prevent the valve assembly from freezing.

This feature ensures that the air dryer continues to provide efficient and reliable operation under extreme climatic conditions.

Fig. 11: Air dryer

Air dryer with integral pressure regulator

Design

Like the unit described above, the air dryer with integral pressure regulator consists of the control stage (components 9...14) and the dessicant cartridge (1). Also mounted on this unit is a pressure regulator (components 4...8, Fig. 12). This regulator controls internal pressure within the brake system. It also controls the air dryer by switching to the drying mode when the system's pressure is below the cut-off point, and activating the regeneration mode when the system reaches cut-in pressure – in other words, in the idle state.

Operation

Drying process

The humid compressed air flows through connection 1 and into the dessicant cartridge (1), where it dries while flowing through the dessicant (3) before continuing through connection 21, leading to the air tanks, or connection 22, which discharges into the regeneration reservoir.

Although the pressure at connection 21 is exerted against the regulator's diaphragm (7), it is not sufficient to compress the spring (8) toward the left as long as the system remains below the cut-off pressure. The passage between connection 21 and the chamber above the bleeder valve (10) is closed, and the bleeder valve remains closed.

Regeneration

When the pressure at connection 21 reaches the cut-off pressure, it overcomes the resistance exerted by the spring (8) and pushes the diaphragm (7) in the pressure regulator to the left. The passage between connection 21 and the chamber above the bleeder valve (10) is now open. The bleeder valve responds by opening the passage between connection 1 and discharge 3 on the drain fitting (11), allowing the air from the compressor to escape to the atmosphere.

During this process, a passage is also opened between this discharge valve and the dessicant cartridge (1), which depressurizes in response. Dried compressed air flows from the regeneration reservoir and through connection 22 before emer-

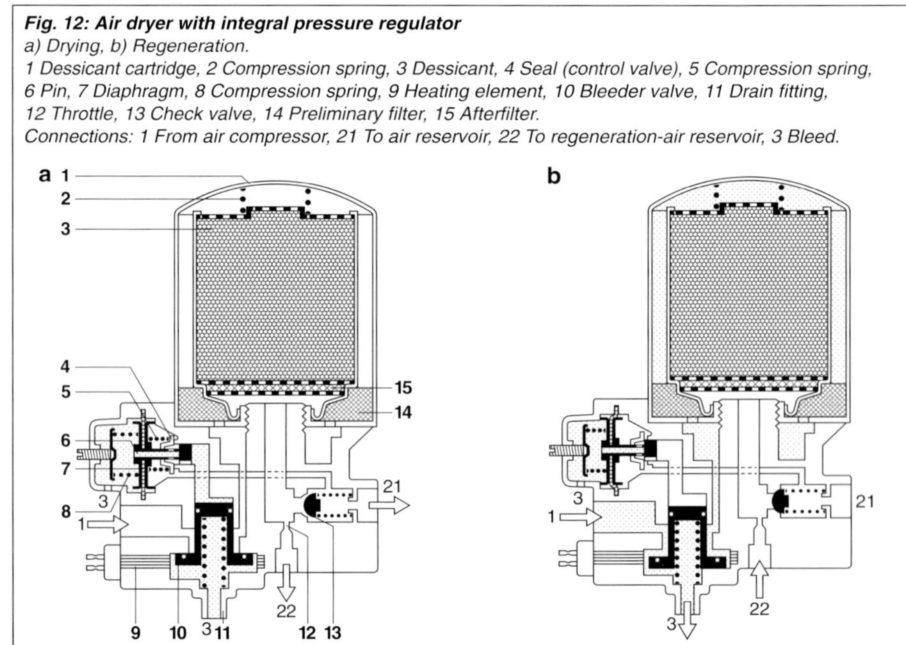

Fig. 12: Air dryer with integral pressure regulator
a) Drying, b) Regeneration.
1 Dessicant cartridge, 2 Compression spring, 3 Dessicant, 4 Seal (control valve), 5 Compression spring, 6 Pin, 7 Diaphragm, 8 Compression spring, 9 Heating element, 10 Bleeder valve, 11 Drain fitting, 12 Throttle, 13 Check valve, 14 Preliminary filter, 15 Afterfilter.
Connections: 1 From air compressor, 21 To air reservoir, 22 To regeneration-air reservoir, 3 Bleed.

ging from the throttle (12). The air current then expands to atmospheric pressure before continuing on through the dessicant (3), where it absorbs moisture. This process regenerates the dessicant, preparing it for drying the next charge of humid air as soon as the pressure regulator switches again.

Air-pressure sensor

Function

This contact-type switch triggers an optical or acoustic alarm signal whenever the system pressure drops below the defined minimum level. The switch is designed as a normally closed (n.c.) switch, in which the contacts remain open during normal operation, and close to activate the low-pressure alarm (Fig. 13).

Design

A diaphragm (6) separates the pressurized area (10) in the housing (7) from the contact area (8) in the insulator (2). The two contacts – a mobile disk (4) on the bottom and a stationary pressure plate (5) at the top, open and close to control the electrical circuit between the blade terminal (1) and the housing (7). A compression spring (3) exerts pressure against the mobile contact disk from above.

Operation

Normal operating mode
When the system pressure is at normal levels, the compressed air exerts force against the diaphragm (6) and the pressure spindle (9) to compress the spring (3) and lift the lower contact disk (4) from the pressure plate (5). This interrupts the flow of electrical current between the blade terminal (1) and the housing (7).

Alarm mode
When the system pressure drops below the specified minimum, the compression spring (3) presses the lower contact disk (4) against the pressure plate (5), allowing electrical current to flow between the blade terminal (1) and the housing (7). In this state the air-pressure switch closes the circuit, triggering an optical or acoustic alarm signal.

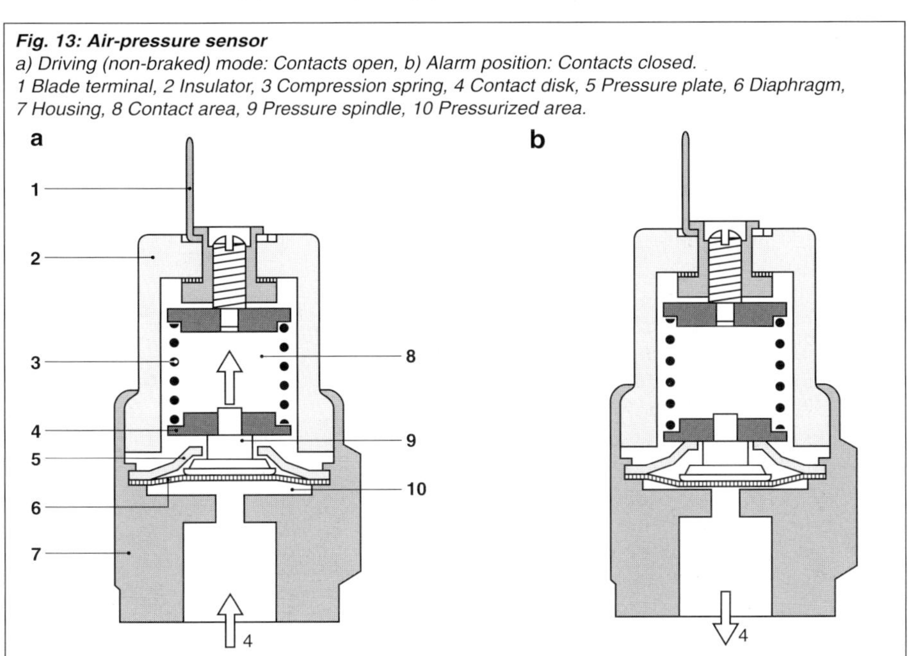

Fig. 13: Air-pressure sensor
a) Driving (non-braked) mode: Contacts open, b) Alarm position: Contacts closed.
1 Blade terminal, 2 Insulator, 3 Compression spring, 4 Contact disk, 5 Pressure plate, 6 Diaphragm,
7 Housing, 8 Contact area, 9 Pressure spindle, 10 Pressurized area.

Overflow Valve

Purpose

Overflow valves safeguard compressed-air loads or compressed-air circuits with respect to one another so that, in the event of a loss of pressure in one circuit, the other circuit continues to be supplied with the compressed air.

Dual-circuit protection valves and 4-circuit protection valves are constructed using two or four overflow valves, respectively. In order to understand these two components, it is necessary to understand the design and operating principles of the various types of overflow valves.

Types
The following types of overflow valves are available (Figs. 14 and 15):
– With full return flow
– With limited return flow
– Without return flow.

Full return flow between two air reservoirs is used, for example, in order to ensure immediate braking readiness of the system and, in addition, to achieve the 9 brake actuations with a stopped air compressor required according to RREG. Limited return flow is used if the pressure to the loads may drop by a certain amount if the first circuit fails, e.g. for exhaust brakes and other secondary loads. Overflow valves without return flow are used if, in the case of two secondary load circuits, one circuit fails and the pressure is to be maintained in the other circuit, for example, for pneumatic suspension.

Design

A helical compression spring (6) presses the diaphragm (2) onto the valve seat (3). In order to obtain an opening pressure independent of the secondary pressure, the valve diameter is equal to the inner diameter (effective diameter) of the diaphragm. The upper side of the diaphragm is at atmospheric pressure, i.e. the overflow valve is not under pressure.

RREG Directives of the Council of the European Community.

The diaphragm opening pressure can be set within limits using the adjusting screw (1). The diaphragm's "breathing space" is bled by means of the check valve (5). This check valve prevents the entry of water and dirt. Overflow valves with full return flow and without return flow differ in the effective direction of action of the check valve (4) which is not provided in overflow valves with limited return flow.

Operating principle

When pressureless, there is no connection between port 1 (primary side) and port 2 (secondary side).

If the compressed air is fed to port 1, a pressure (the primary pressure) builds up under the valve seat (3). This pressure opposes the force of the compression spring (6) and lifts the diaphragm (2) off of the valve seat as soon as the force of the air exceeds that of the spring, i.e. when the opening pressure has been achieved. The compressed air can flow from port 1 to port 2, and the secondary pressure rises to the level of the primary pressure. If the pressure at port 1 drops, the compressed air of the secondary side can flow back from port 2 to port 1 either fully, to a limited extent or not at all depending upon the valve version.

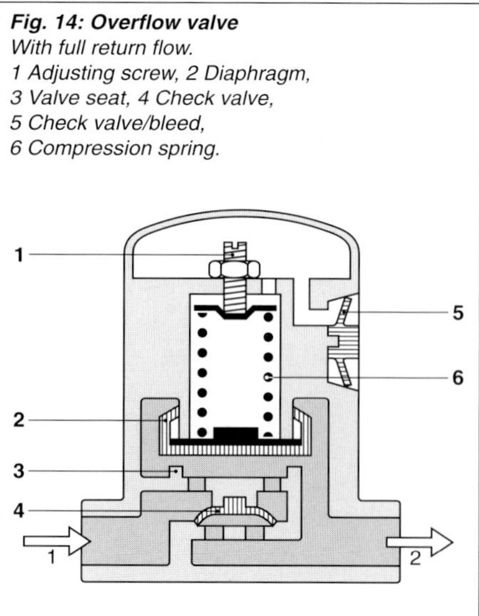

Fig. 14: Overflow valve
With full return flow.
1 Adjusting screw, 2 Diaphragm,
3 Valve seat, 4 Check valve,
5 Check valve/bleed,
6 Compression spring.

Overflow valve with full return flow

Compressed air flows from port 2 through the opened valve to port 1 until the pressure on the secondary side has dropped to the closing pressure (= opening pressure). In doing so, the diaphgram (2) is forced up against the valve seat (3) because the force of the spring (6) is greater than that of the air. As the primary pressure continues to fall, the compressed air continues flowing from port 2 to port 1 but now by way of the open check valve (4). For this reason, the pressure can drop to atmospheric at both ports (Fig. 15a).

Overflow valve with limited return flow

The compressed air can only flow back from port 2 to port 1 as long as the diaphragm (2) is not up against the valve seat (3) (Fig. 15a). That is, only when the pressure is still above the closing pressure.

Overflow valve without return flow

The check valve (4) prevents any return flow of compressed air from port 2 to port 1. If the pressure on the primary side drops, the secondary pressure remains unchanged (Fig. 15b).

Dual-circuit protection valve

Purpose

The dual-circuit protection valve has a three-fold purpose in dual-circuit compressed-air systems:
- Distribution of the compressed air from one energy source to two supply circuits separated from one another
- Safeguarding the pressure in the intact circuit in the event of the failure of one of the two circuits
- Safeguarding the pressure in both circuits in the event of the failure of the energy source.

Design

The dual-circuit protection valve consists of two overflow valves. When the opening pressure has been reached, these valves allow compressed air to flow from the energy source to the air reservoirs and permit the compressed air to flow back either not at all or only up to a certain closing pressure. Both overflow valves are supplied with compressed air by way of the common port 1. They are connected to the two supply circuits via the ports 21 and 22. Depending upon the type of return flow, one distinguishes between two different versions:
- Both circuits with limited return flow

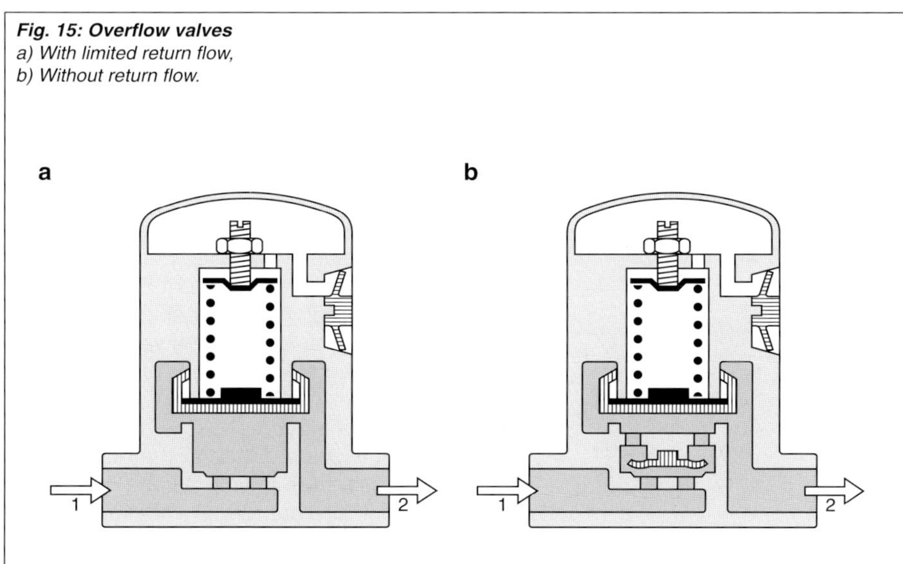

Fig. 15: Overflow valves
a) With limited return flow,
b) Without return flow.

– One circuit with limited return flow, the other without return flow (for safeguarding high pressure, e.g. in the case of spring-type brake actuators).

Limited return flow in at least one of the two overflow valves is necessary so that, as the compressed air is used, information on the drop in pressure is transmitted to the pressure regulator in order that this can switch to supplying air again as soon as the cut-in pressure has been reached.

Operating principle
The pressure regulator and the dual-circuit protection valve work together closely, and their control pressures are matched to each other. The pressure regulator's cut-off pressure must be above the opening pressure of the dual-circuit protection valve. The opening pressure is defined as that pressure which is necessary to open the dual-circuit protection valve when the system is empty. When the system is full, this pressure is somewhat lower because the supply pressure at the ports 21 and 22 acts on the diaphragm (2).

The pressure-regulator cut-in pressure must be above the static closing pressure of the dual-circuit protection valve. If, for example, a considerable loss of pressure occurs in circuit 21 due to a leak, the pressure regulator switches to the filling position before the dual-circuit protection valve closes port 21.

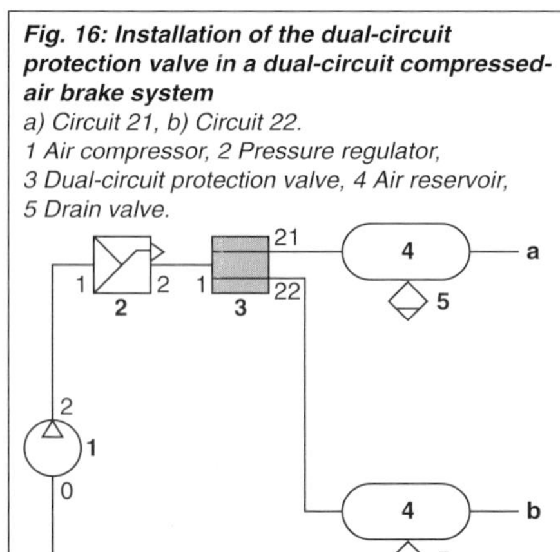

Fig. 16: Installation of the dual-circuit protection valve in a dual-circuit compressed-air brake system
a) Circuit 21, b) Circuit 22.
1 Air compressor, 2 Pressure regulator,
3 Dual-circuit protection valve, 4 Air reservoir,
5 Drain valve.

Dual-circuit protection valve

Circuits 21 and 22 with limited return flow

Intact system
When the compressed-air braking system is empty, the helical compression springs (1) press the valve plates (3) onto the valve seats (4).

The compressed air flowing through port 1 causes pressure build-up beneath the valve plates. If the opening pressure is reached, the valve plates lift off of the valve seats and the compressed air flows into the air reservoirs by way of the ports 21 and 22. When the cut-off pressure of the pressure regulator is reached in the air reservoirs, the pressure regulator switches to "idle" and the compressor feeds the compressed air to atmosphere. The pressure in both supply circuits is now higher than the closing pressure. The force of the air beneath the diaphragm (2) exceeds that of the helical compression springs so that the valve plates remain off of the valve seats. As compressed air is used, the pressure decreases until the pressure-regulator cut-in pressure is reached so that compressed air is again fed into the air reservoirs before both valve plates in the dual-circuit protection valve can close (Fig. 17a).

Failure of a circuit
If one of the circuits starts to leak and the pressure in this circuit's air reservoir drops to the closing pressure of the dual-circuit protection valve, the valve plate (3) is closed by the force of the spring (1). Prior to this, since its cut-in pressure is reached before the closing pressure of the dual-circuit protection valve, the pressure regulator switches to the filling mode.

The opening pressure in the defective circuit is greater than in the intact circuit because there is no pressure to assist in lifting the diaphragm. This means that compressed air is fed into the intact circuit until the defective circuit's opening pressure has been reached. If the leak in the defective brake circuit is so severe that compressed air escapes at a rate faster than it can be supplied, the pressure regulator cut-off pressure can no longer be reached and the air compressor constantly supplies compressed air to the dual-circuit protection valve. The operating pressure in the intact circuit then corresponds to the opening pressure of the defective circuit.

Dual-circuit protection valve

Circuit 21 with limited return flow, circuit 22 without return flow

Intact system
The overflow valve with limited return flow operates in the same manner as the overflow valves described in the previous section. The overflow valve without re-

turn flow acts like a check valve when the pressure regulator is switched to the idle mode, and prevents compressed air from returning from circuit 22 to the energy source. The valve plate (3) can only open if the opening pressure has been exceeded.

If the force of the air in circuit 22 is less than that of the spring (1), the spring presses the valve plate (3) against the valve seat (4) by way of the diaphragm (2) and the guide pin (6).

If the force of the air in circuit 22 exceeds that force of spring (1), the valve plate is forced against the valve seat by spring (5) when the pressure regulator is switched to the idle mode (Fig. 17b).

Failure of a circuit
In principle, the reactions in the event of one of the circuits failing are the same as those for the dual-circuit protection valve with limited return flow.

However, no compresed air flows back from circuit 22 to port 1.

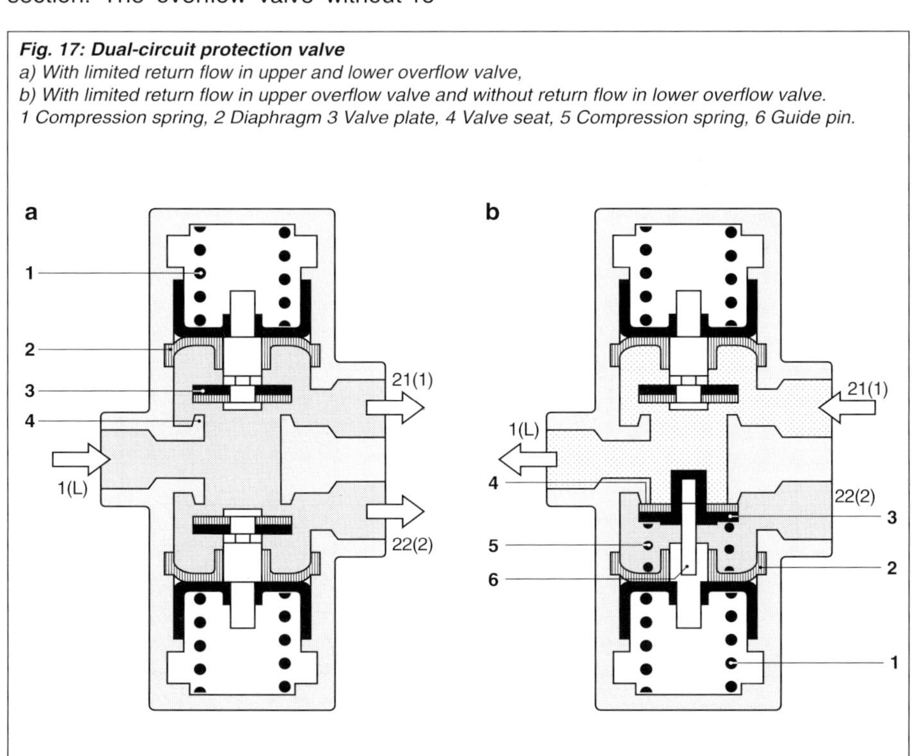

Fig. 17: Dual-circuit protection valve
a) With limited return flow in upper and lower overflow valve,
b) With limited return flow in upper overflow valve and without return flow in lower overflow valve.
1 Compression spring, 2 Diaphragm 3 Valve plate, 4 Valve seat, 5 Compression spring, 6 Guide pin.

Four-circuit protection valve

Purpose

The four-circuit protection valve is used in dual-circuit compressed-air braking systems for supplying compressed air and for safeguarding the pressure in
- the two service-brake circuits,
- the parking-brake circuit,
- the trailer circuit, and
- the secondary load circuit.

Design

Four overflow valves with limited-return flow are arranged in a single housing. They can be triggered either sequentially in pairs, or they can be centrally triggered simultaneously. The protection valve with the circuits arranged in pairs for sequential triggering will be dealt with in the following text.

Ports 21 and 22 are assigned to the overflow valves I and II (Fig.18).

The overflow valves III and IV are connected downstream of the overflow valves I and II by way of check valves. The secondary side of overflow valves III and IV is routed to ports 23 and 24.

Operating principle

Intact system

With the air reservoirs empty, in all four valves I-IV, the helical compression spring (2) presses the diaphragm piston (3) onto the valve seat (4). The valve is closed. As soon as the pressure of the compressed air flowing into port 1 is equal to the valve opening pressure, valves I and II open. However, it is possible that one of these two valves opens before the other. Compressed air now flows via ports 21 and 22 into the circuits 21 and 22 where it builds up a pressure.

This pressure also counteracts the force of the compression spring (2) since the total effective area of the diaphragm piston (3) comes into effect.

An increasing portion of the compression-spring force is absorbed by the increasing force of the air so that the minimum pressure necessary to keep the valve open is lower than the opening pressure. The pressure in the air reservoirs equals the pressure of the compressed air flowing in at port 1.

As the pressure continues to rise, the diaphragm piston lifts further away from

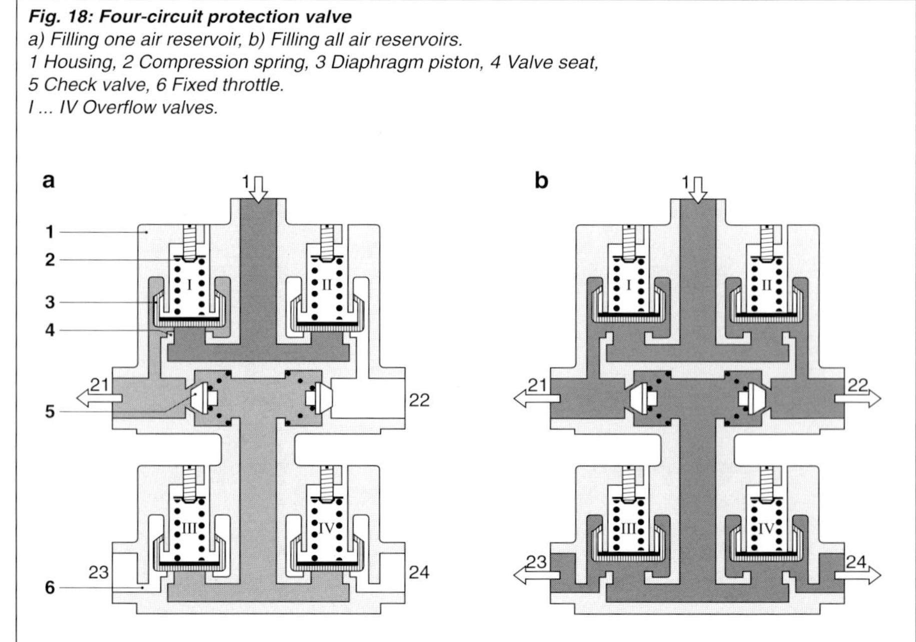

Fig. 18: Four-circuit protection valve
a) Filling one air reservoir, b) Filling all air reservoirs.
1 Housing, 2 Compression spring, 3 Diaphragm piston, 4 Valve seat,
5 Check valve, 6 Fixed throttle.
I ... IV Overflow valves.

the valve seat, whereby the complete flow cross-section of the valve seat is opened.

As a result of the valve plate having lifted from the valve seat (due to the small cross-section), the pressure necessary for opening the valve is greater than the theoretical opening pressure which can be calculated using the valve-seat area and the spring force.

The air reservoirs of circuits 21 and 22 are filled either simultaneously or one after the other, depending upon whether the opening pressures of the overflow valves I and II are identical or whether they differ slightly from one another. The compressed air flows to the overflow valves III and IV by way of the two check valves (5) which safeguard the circuits 21 and 22 from one another.

These overflow valves are set to the same opening pressures as overflow valves I and II, and operate in the same manner. In addition, fixed throttles (6) are located downstream of these overflow valves. In the event of a high rate of air consumption, such as that occurring during the filling of spring-type brake actuators in two-reservoir systems, at port 23 for instance, these fixed throttles prevent the pressure beneath the diaphragm piston (3) dropping too far. They also ensure that the valve either closes very rapidly or at a higher pressure than desired. Due to the throttling effect, sufficient compressed air can be supplied from other circuits in the event of a high rate of compressed-air consumption in one circuit.

Defective system

General
As soon as a leak occurs in the braking system which is so large that the delivery capacity of the air compressor can no longer compensate for the loss, the four-circuit protection valve safeguards the circuits which are still intact. The pressure at which the defective circuit's overflow valve closes, depends upon the magnitude of the compressed-air loss.

If the pressure in the defective circuit drops very slowly, the pressure beneath the total effective area of the diaphragm piston (3) is uniform (static closing pressure). If the leak is larger, a throttling effect occurs at the valve seat between it and the diaphragm piston. As a result, the pressure beneath the valve-seat ring area drops at a greater rate than it does beneath the valve-seat area and thus relieves the load on the helical compression spring (2) so that this closes the valve at a higher pressure (dynamic closing pressure). In the event of a break in a line, the dynamic closing pressure reaches the level of the opening pressure.

Failure of a service-brake circuit
(Circuit 21 or 22)
The pressure in the intact service-brake circuit drops until the diaphragm piston (3) closes the valve seat in the defective circuit and thus isolates it. The check valves (5) prevent a drop in pressure in the circuits 23 and 24. The air compressor now fills the intact circuit until the opening pressure of the defective circuit is reached.

Failure of circuit 23 or 24
Due to the effects of continued air flow, the pressure in the intact circuits drops to the closing pressure of the defective circuit. Finally, the pressure rises until the defective-circuit opening pressure is reached.

All supply reservoirs are empty,
one service-brake circuit has failed
Even in the event of unfavorable tolerances in the opening pressure, the intact service-brake circuit is filled. As soon as a slight pressure has been built up in one of the intact service-brake circuits, the overflow pressure of the corresponding overflow valve drops so that the entire delivery capacity flows into the intact brake circuit by way of the completely opened valve. Circuits 23 and 24 are filled in the same manner until finally all intact circuits have reached the opening pressure of the defective circuit.

Drain valve

Purpose

The drain valve is used for draining condensate from the air reservoirs of compressed-air systems.

On those systems not equipped with an air dryer, this is necessary in order to prevent the condensate level in the air reservoirs rising to such a level that the compressed-air volume available to the compressed-air system is reduced so far that the braking efficiency is adversely affected.

In the case of compressed-air systems with air dryer, the drain valve is used to check the correct functioning of the air dryer.

Design

The helical compression spring (4) presses the valve plate (1) against the valve seat (2) and closes the valve. The valve seat is permanently connected to the valve tappet (3) (Fig. 19).

Operating principle

By pulling the ring (5) horizontally in any direction, the valve tappet (3) tilts and lifts the valve plate (1) away from the valve seat (2). The condensate can now drain out of the air reservoir.

Fig. 19: Drain valve
1 Valve plate, 2 Valve seat, 3 Valve tappet, 4 Compression spring, 5 Pull ring.

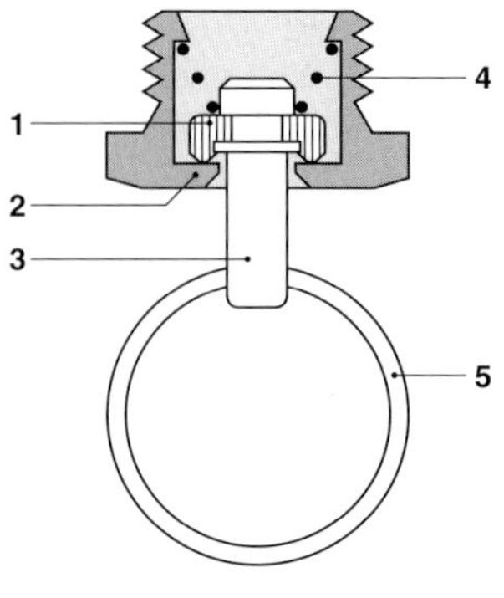

Pressure relief element

Purpose

The pressure relief element is used to limit the pressure in the system to a specific level, i.e. to reduce an incoming supply pressure (primary pressure) to a specific pressure level (secondary pressure).

Limiting the pressure is necessary, for example, in the following cases:

– In systems with high-pressure storage and low-pressure operation for all loads

– In systems with high-pressure pneumatic suspension and low-pressure braking system.

Design

The housing (1) contains a working piston (6) with a sealing cone (3). When the set maximum pressure is reached on the secondary side (port 2), the cone is pressed against the seal ring (4) thus blocking the supply of air from the primary side (port 1).

Sealing cone and seal ring form the border between the primary and secondary sides. The load on the helical compression spring (2) for the working piston (6) can be set by means of the adjusting screw (9), thus setting the level of the secondary pressure (Fig. 20).

The breathing space inside the working piston is vented by means of a check valve (8).

Operating principle

As long as the set maximum pressure has not been reached on the secondary side (port 2), the helical compression spring (2) presses the working piston (6) up as far as it can go.

Compressed air entering on the primary side (port 1) flows through the gap between the sealing cone (3) and the seal ring (4) to the secondary side (port 2).

Pressure builds up on the secondary side and, after exceeding the force of the helical compression spring (2), pushes the working piston (6) downward until the sealing cone comes up against the seal ring and separates the secondary side from the primary side. This means that compressed air can no longer flow to the secondary side where the set maximum pressure has now been reached In this condition, equilibrium has been reached between the force of the piston and the force of the spring.

In view of the fact that the effective diameter of the sealing cone formed by the sealing ring is equal to the diameter at the lower part of the working piston, the two compressed-air forces acting on the working piston from the primary side cancel each other out.

For this reason, as long as it exceeds the secondary pressure, changing the primary pressure has no effect on the secondary pressure.

If the primary pressure drops to below the secondary pressure, the compressed air flows from port 2 to port 1 by way of the seal lip (5) on the seal ring (4) which has been pushed away by the higher secondary pressure. In the opposite direction the seal lip (5) acts as a check valve. In this manner, a component connected on the secondary side, e.g. a brake cylinder, can be vented by reducing the primary pressure.

If the secondary pressure drops slightly (≤0.1 bar), the working piston (6) opens the passage to the secondary side immediately until equilibrium of the forces has been reached again and the secondary pressure has again reached the set level.

Increasing the initial tension of the spring (2) by means of the adjusting screw results in an increased secondary pressure, reducing the initial tension in a reduced secondary pressure.

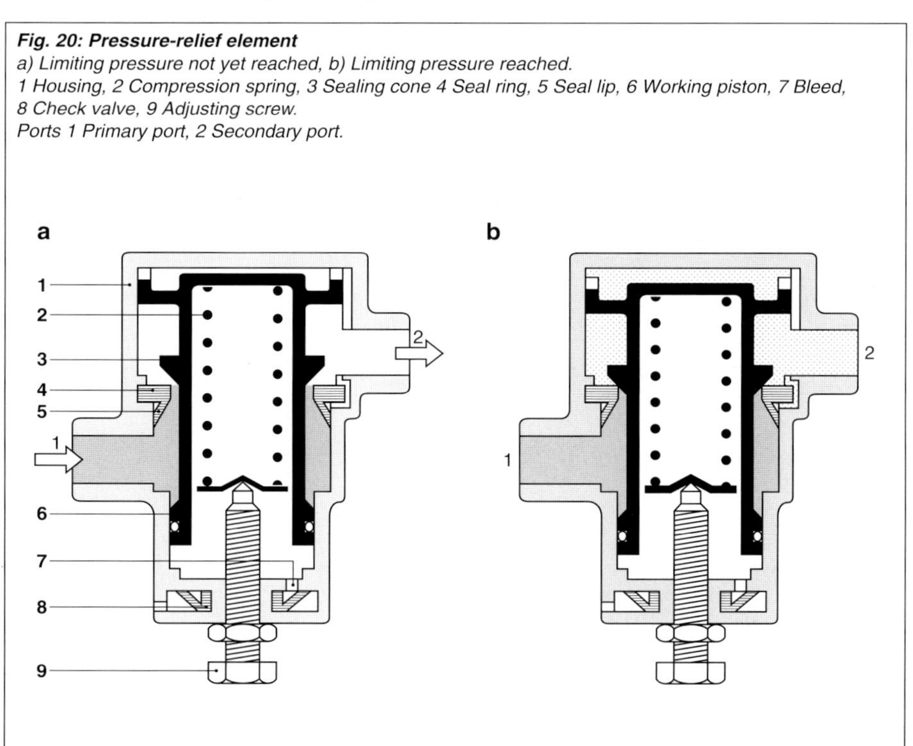

Fig. 20: Pressure-relief element
a) Limiting pressure not yet reached, b) Limiting pressure reached.
1 Housing, 2 Compression spring, 3 Sealing cone 4 Seal ring, 5 Seal lip, 6 Working piston, 7 Bleed, 8 Check valve, 9 Adjusting screw.
Ports 1 Primary port, 2 Secondary port.

Component group B "Service-brake system"

Purpose and design

To reduce the vehicle's speed, to keep it constant on a downgrade, to stop the vehicle (Fig. 1).

In medium-heavy and heavy-duty commercial vehicles, the pedal force applied by the driver is insufficient on its own for achieving adequate braking deceleration. Usually, in such cases compressed-air power-brake systems are used, whereby the compressed air is used as the stored energy for actuating the service-brake system.

Depending upon the size of the vehicle, there are a number of different designs for the service-brake system. Basically speaking, there are four different types of service-brake systems. The design of each is shown in the following four diagrams (Figs. 2 and 3)

The corresponding components are described on the following pages.

Fig. 1: Component group B: "Service-brake system"

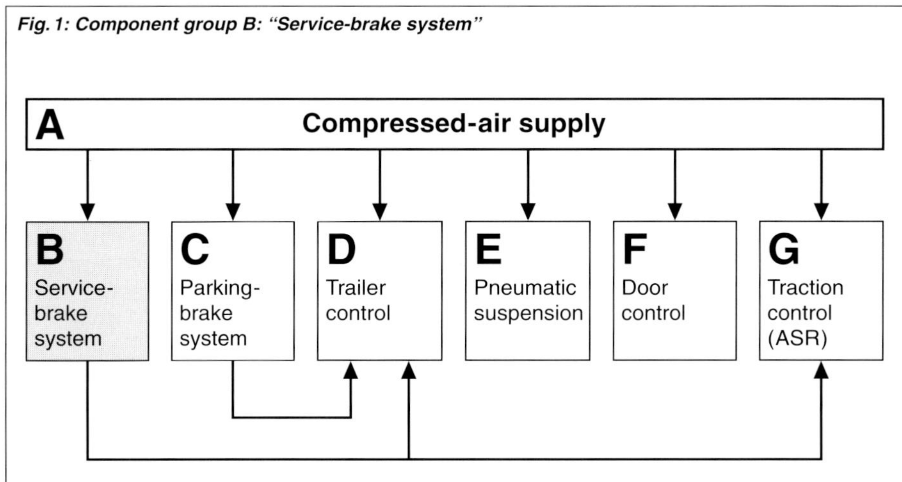

Fig. 2: Single-circuit energy-assisted compressed-air brake system with hydraulic transmission device

Components used:

1 Brake servo-unit cylinder, 2 Tandem master cylinder, 3 Brake wheel cylinder.

VA Front axle, HA Rear axle.

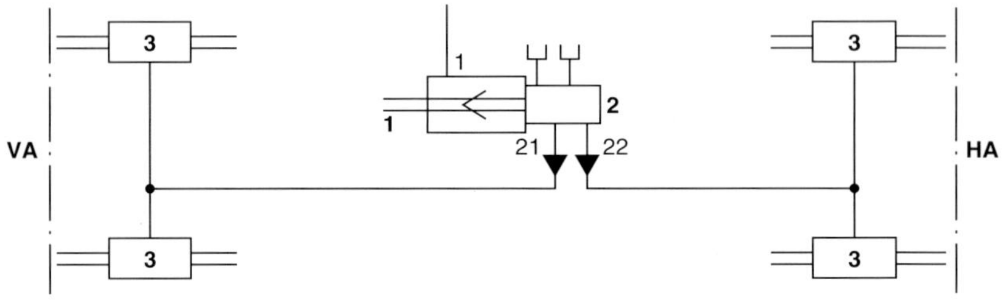

Fig. 3: Dual-circuit compressed-air power-brake system
a) With hydraulic transmission device. Components used:
1 Dual-circuit brake assembly, 2 Tandem master cylinder, 3 Brake wheel cylinder.
VA *Front axle,* HA *Rear axle.*

b) With hydraulic transmission device. Components used:
1 Service-brake valve, 2 Dual-circuit actuator cylinder for brake master cylinder, 3 Tandem master cylinder,
4 Brake wheel cylinder.
VA *Front axle,* HA *Rear axle.*

c) With pneumatic transmission device. Components used:
1 Service-brake valve, 2 Brake wheel cylinder, 3 Combi-brake cylinder.
VA *Front axle,* HA *Rear axle.*

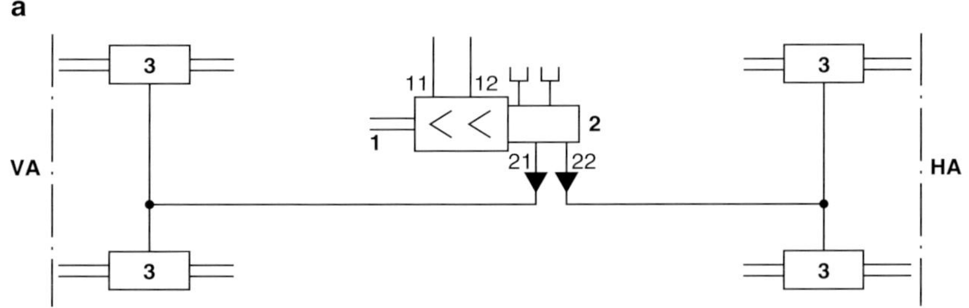

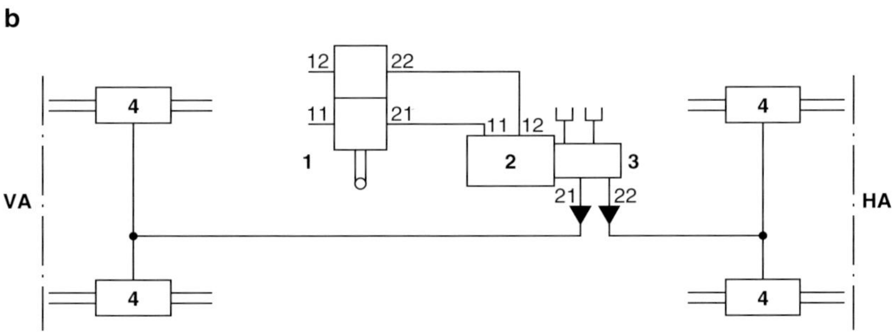

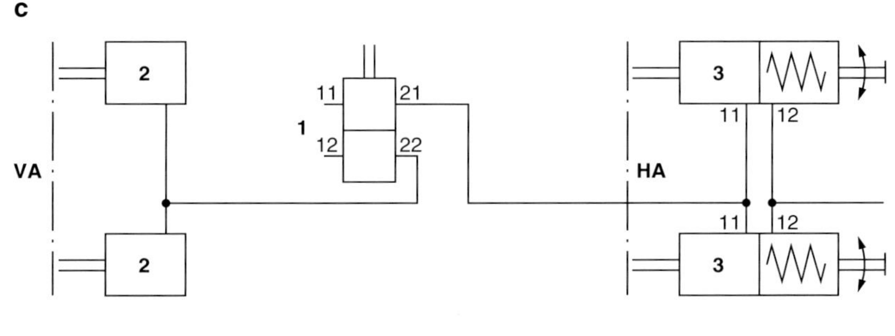

Brake servo-unit cylinder

Purpose

The brake servo-unit cylinder controls the braking energy for the service-brake system in light commercial vehicles. Here, the pedal force supplied by the driver is amplified by means of compressed air and transmitted to the wheel-brake cylinders by means of a hydraulic master cylinder (tandem master cylinder). In the event of a failure of the compressed-air system, the hydraulic master cylinder is actuated mechanically.

Design

The push rod (1), valve (5, 6, 7, 12, 13), servo-unit piston (3) and spindle (10) are arranged centrally one behind the other in the piston housing (14). The push rod (1) is connected to the brake pedal and transmits the driver's pedal force to the servo-unit piston (3). The valve (5, 6, 7, 12, 13) is located in the servo-unit piston (3) and controls the compressed air which amplifies the pedal force inputted by the driver. The servo-unit piston (3) acts on the spindle (10) which actuates the hydraulic master cylinder screwed to the housing floor. The servo-unit piston (3) separates the piston housing (14) into chamber I (pressurized area) and chamber II (pressureless area). The diameter of the piston and the selected operating pressure determine the amplification ratio (Fig. 4).

In contrast to the actuating mechanisms described in the following chapters, the brake servo-unit cylinder is connected only to one compressed-air supply circuit. Nevertheless, in the event of a failure of the compressed-air system, the service-brake system can provide residual braking because the brake servo-unit cylinder in this case actuates the dual-circuit hydraulic master cylinder in the purely mechanical fashion using pedal force.

Operating principle

Driving (non-braked) mode

When the brakes are not applied (driving mode), the plunger spring (9) keeps the servo-unit piston (3) in the basic position. The compressed air transmitted by the pressure hose (8) – assisted by the valve spring (7) – forces the valve plate (6) onto the valve seat (13). The plunger return spring (4) keeps the valve tappet (11) in the "off" position, and the valve seat (12) is lifted off of the valve plate (6). Chamber I (pressurized area) is bled by way of the valve tappet (11).

Partial braking

With the brakes partially applied, i.e. when the brake pedal is not pushed all the way to the floor, the push rod (1) presses the valve tappet (11) with the valve seat (12) against the valve plate (6) so that chamber I

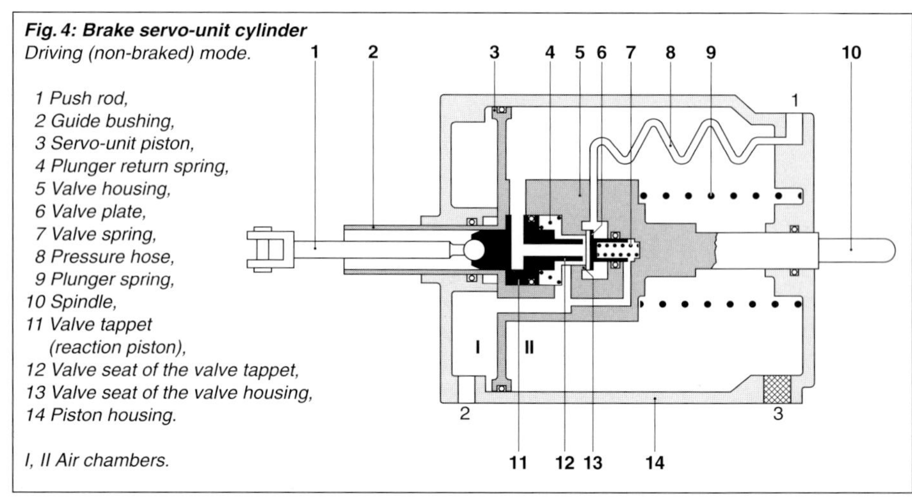

Fig. 4: Brake servo-unit cylinder
Driving (non-braked) mode.

1 Push rod,
2 Guide bushing,
3 Servo-unit piston,
4 Plunger return spring,
5 Valve housing,
6 Valve plate,
7 Valve spring,
8 Pressure hose,
9 Plunger spring,
10 Spindle,
11 Valve tappet
(reaction piston),
12 Valve seat of the valve tappet,
13 Valve seat of the valve housing,
14 Piston housing.

I, II Air chambers.

(the pressurized area) is no longer bled. At the same time, the valve plate (6) is lifted off of the valve seat (13). This causes compressed air to flow into chamber I (the pressurized area). Due to the fact that the vent of the pressurized area is blocked, the incoming compressed air causes a rise in pressure which is applied to the servo-unit piston (3) and – in the opposite direction – to the valve tappet (11) which acts as a reaction piston during partial braking. If, in the partial-braking mode, equilibrium is now achieved between the pedal force pressing on the valve tappet (11) and the pressure exerted by the compressed air and plunger return spring (4) in the opposite direction on the valve tappet (reaction piston), the valve plate (6) closes the valve seat (13). Valve seats (12) and (13) are now both closed at the same time. If the position of the brake pedal remains unchanged, a partial pressure remains in chamber I (pressurized area) because compressed air can no longer reach the servo-unit piston but also cannot escape from the pressurized area.

Fully braked mode

If the brake pedal is pushed all the way to the floor, the push rod (1) presses valve tappet (11), complete with valve seat (12), against valve plate (6). The vent of chamber I (pressurized area) is now blocked. At the same time, valve plate (6) is lifted away from valve seat (13) so that compressed air from pressure hose (8) flows into chamber I (pressurized area) and presses against the servo-unit piston (3). Assisted by pedal force and compressed air, the servo-unit piston (3) moves to the right as shown Fig. 5b, and pushes the spindle (10) into the hydraulic master cylinder until this stops the motion of the servo-unit piston (3). The supply pressure and the brake pressure are at the same level.

Failure of the compressed-air system

If the compressed-air system fails, the spindle (10) can be actuated mechanically through the brake servo-unit, although greater pedal force must be applied. In this case, the push rod

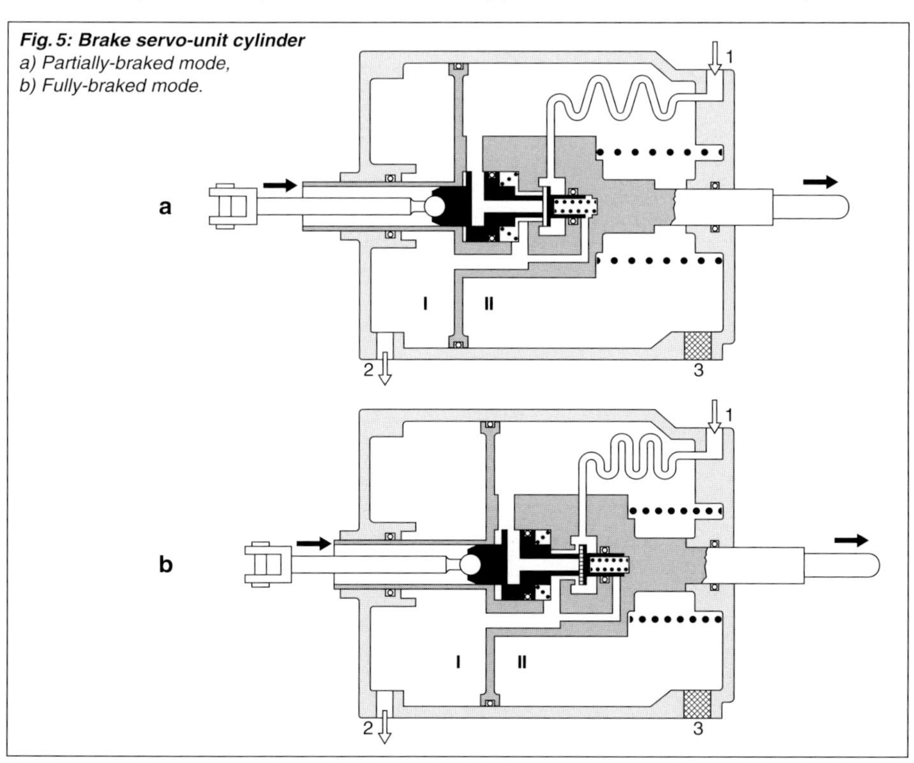

Fig. 5: Brake servo-unit cylinder
a) Partially-braked mode,
b) Fully-braked mode.

presses the valve tappet (11) all the way to the valve housing. The valve housing (5) then shifts the servo-unit piston (3) against the force of the plunger spring (9), so that the spindle (10) is pressed into the hydraulic master cylinder.

Dual-circuit brake assembly

Purpose
The dual-circuit brake assembly is used for controlling and actuating the hydraulic transmission devices by means of compressed air. In accordance with brake-pedal position, compressed air flows into the brake assembly and acts on the pistons which mechanically actuate the flanged-on hydraulic tandem master cylinder by way of the spindle.

Design
The dual-circuit brake assembly consists of a combination of a dual-circuit drag-piston actuator cylinder for the brake master cylinder and a dual-circuit servo-valve. The servo-valve and the actuator cylinder are arranged axially one behind the other.

The dual-circuit brake assembly is controlled hydraulically or mechanically by pressing the brake pedal. The dual-circuit brake assembly has 2 supply ports 11 and 12, as well as 2 ports for the con-

trol lines 21 and 22 to the trailer-control valve.

Even if only one intact compressed-air brake circuit is available, full braking force is maintained, because of the design of the two-circuit drag-piston actuator cylinder for the brake master cylinder.

Operating principle

Driving (non-braked) mode
When the brakes are not applied, the two valve plates (8, 11) of brake circuits 1 and 2 are up against their respective inlet-valve seats (7, 12). As a result, the actuator cylinder for the brake master cylinder (18) and the two brake lines to the trailer-control valve (ports 21 and 22) are pressureless (Fig. 6).

Partial braking
During partial braking, the brake pedal, which is only partially depressed, pushes the control plunger (3) complete with the discharge-valve seat (6) against the valve plate (8) by way of the tappet (1) and the travel-limiting spring (2). As a result, the valve plate is lifted away from the inlet-valve seat (7).

At first, the outlet to the vent (21) is closed on the control valve of brake circuit 1 before the inlet for the compressed air opens. The incoming compressed air acts on 3 components of the dual-circuit brake assembly.

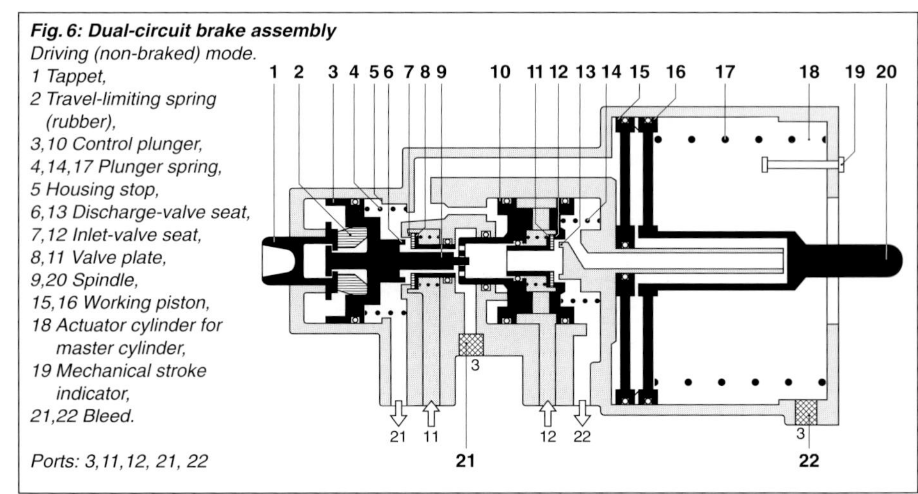

Fig. 6: Dual-circuit brake assembly
Driving (non-braked) mode.
1 Tappet,
*2 Travel-limiting spring
(rubber),*
3,10 Control plunger,
4,14,17 Plunger spring,
5 Housing stop,
6,13 Discharge-valve seat,
7,12 Inlet-valve seat,
8,11 Valve plate,
9,20 Spindle,
15,16 Working piston,
*18 Actuator cylinder for
master cylinder,*
*19 Mechanical stroke
indicator,*
21,22 Bleed.

Ports: 3,11,12,21,22

– Control plunger (3) of brake circuit 1
– Control plunger (10) of brake circuit 2
– Working piston (15).

The compressed air builds up a brake pressure (partial pressure) at the control plunger (3) of brake circuit 1 until this return force, together with the force of the plunger spring (4), has exceeded the actuation force transmitted by the brake pedal and brought the control plunger back to the center position. In this position, the inlet and outlet of the control valve are closed. The more the brake pedal is depressed, the higher the brake pressure. The distance travelled by the control plunger in the direction of the tappet is absorbed by the travel limiting spring (2) which is compressed in a corresponding manner.

The brake pressure of the brake circuit 1 acts on the control plunger (10) of brake circuit 2 and shifts it against the plunger spring (14). Due to this motion, the control plunger (10) first closes the discharge-valve seat (13) using the valve plate (11) and then opens the inlet-valve seat (12) so that compressed air from port 12 can flow into brake circuit 2 and to the working piston (16). As soon as the action force of brake circuit 1 (pressure x effective area) at the control plunger (10) is equal to the reaction force of brake circuit 2 (pressure x effective area + force of the plunger spring [14]), the inlet-valve seat (12) closes whereas the discharge-valve seat (13) remains closed. The control plunger (10) is now in an equilibrium condition between the two brake circuits. Due to the fact that the effective area of the control plunger (10) facing brake circuit 2 is larger than that facing brake circuit 1 and, at the same time, the force of the plunger spring (14) has an assisting effect, the pressure in brake circuit 2 is approximately 1 bar less than that in brake circuit 1 when the control plunger is in the equilibrium condition. This pressure difference is necessary so that both working pistons (15 and 16) are moved every time the brakes are applied.

The pressures of brake circuits 1 and 2 move the two working pistons (15 and

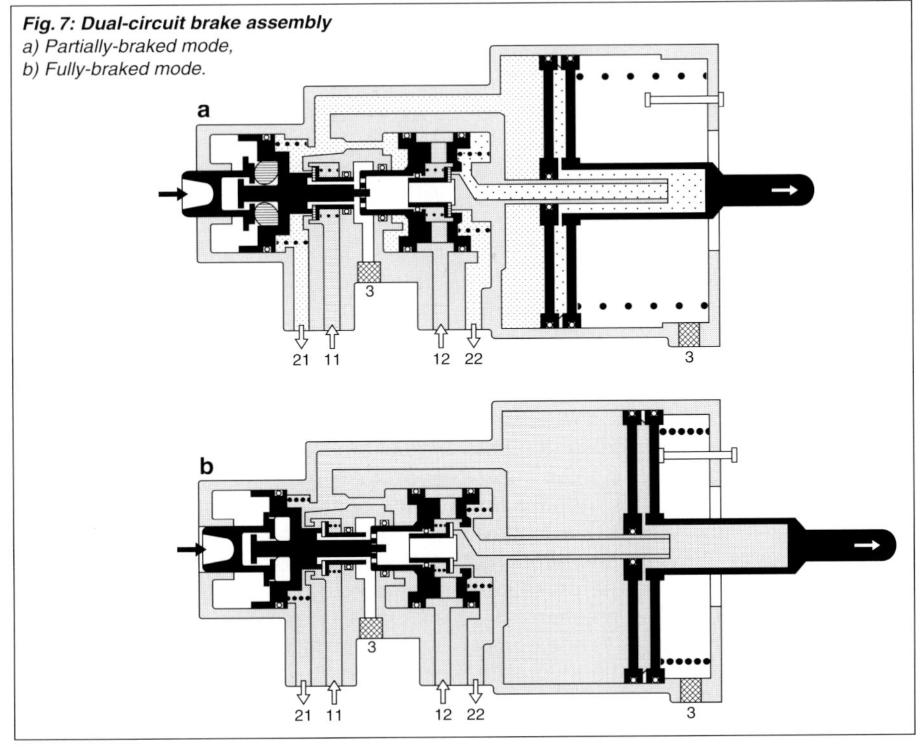

Fig. 7: Dual-circuit brake assembly
a) Partially-braked mode,
b) Fully-braked mode.

16) against the plunger spring (17). The spindle (20) which is connected to working piston (16) transmits the brake force to a hydraulic tandem master cylinder.

Fully braked mode
When the brakes are fully applied, the brake pedal and thus also the tappet (1) are fully depressed. The control plunger (3) of brake circuit 1 then comes up against the housing stop (5). In doing so, first the discharge-valve seat (6) of brake-circuit 1 control valve closes and then the inlet-valve seat (7) opens completely. There is no center position at which both the inlet and the outlet are closed. In the same manner, the outlet (13) is closed and the inlet (12) opened in brake-circuit 2 control valve. The control plunger (10) of brake circuit 2 is controlled first bythe brake pressure of brake circuit 1 and, towards the end of the stroke, mechanically by the control plunger (3). With the brakes fully applied, the brake pressure in both brake circuits is equal to the supply pressures.

Failure of a brake circuit
If brake circuit 1 fails and becomes pressureless, the control plunger (10) of brake circuit 2 must be actuated mechanically by the spindle (9). In order to achieve the same brake pressure as with the intact system, a larger actuating stroke and a slightly larger actuating force are required. In this case, the compressed air acts only on the working piston (16) which then alone transmits the brake force to the hydraulic tandem master cylinder by way of the spindle (20) while the working piston (15) does not move. If brake circuit 2 fails, the actuating force and actuating stroke do not change with respect to those of the intact component. In the actuator cylinder for the brake master cylinder (18), the compressed air acts only on the working piston (15) which pushes the pressureless working piston (16) mechanically against the plunger spring (17) and transmits the brake force to the hydraulic tandem master cylinder by way of the spindle (20).

Service-brake valve with rocking piston

Purpose
The dual-circuit service-brake valve controls two independent pneumatic brake circuits in the tractor (towing) vehicle and – by way of the trailer-control valve – the trailer's service-brake system.

Design
The component consists of two separate brake valves which are arranged one behind the other and which are actuated by a common control, usually a brake pedal. Equal pressures are achieved in both service-brake circuits by the effect of spring and valve sealing forces of equal magnitude on each of the two control sides of the piston (a rocking piston). (This is known as the rocking-piston principle.)
The brake pressure can be metered delicately in both circuits. Short valve response travel is achieved by using a preloaded spring.
If one circuit fails, full functioning of the other circuit is assured.

Operating principle

Driving (non-braked) mode
When the brakes are not applied (rest position, Fig. 8a), the cup seals (7) and (14) are up against the inlet-valve seats (8) and (13). As a result, compressed air cannot flow into the brake circuits 1 and 2 by way of the ports 21 and 22. The ports 21 and 22 are connected to vent 3 so that both brake circuits are vented.

Partial braking
The brake pedal is partially depressed during partial brake application (Fig. 8b). The tappet (1) pushes the reaction piston (3) downward by way of the travel-limiting spring (2) until the discharge-valve seat (9) is closed. The rocking piston (10) is pushed downward by way of the plunger spring (6) so that the discharge-valve seat (11) also closes and then the inlet-valve seats (8) and (13) open.

The inlet-valve seats remain open until the compressed air entering by way of port 11 has built up sufficient force beneath the reaction piston (3) to push it upward against the force of the travel-limiting spring (2) and close the inlet valve seat (8) again. The inlet and outlet of brake circuit 1 are then closed, and the valve is in the center position.

Together with the reaction piston (3), the rocking piston (10) moves upward and closes the inlet-valve seat (13) so that the brake pressures in brake circuits 1 and 2 are equal.

The magnitude of the pressures in both brake circuits is a function of the actuating force. The more the brake pedal is depressed, the higher is the brake pressure.

Fully braked mode

During full brake application, the brake pedal is pressed all the way to the floor (Fig. 8c). The tappet (1) is pressed down to such an extent that, after overcoming the force of the travel-limiting spring (2), the reaction piston (3) is pushed downward until it reaches the stop (5). The rocking piston (10) is also pressed downwards by way of the helical compression springs (4) and (6) until it reaches the stop (12). During the downward motion of these two pistons, the two discharge-valve seats (9) and (11) close first, then the two inlet-valve seats (8) and (13) open and remain open as long as the brake pedal is completely depressed, the brake pressure in both brake circuits is equal to the supply pressures.

Failure of a brake circuit

If a brake circuit fails, the brake pressure in the intact brake circuit can still be precisely metered. If brake circuit 1 (port 11) fails so that the piston (3) cannot act as the reaction piston during partial brake application, the rocking piston (10) performs the function of a reaction piston and closes the inlet-valve seat (13) as soon as sufficient pressure has built up beneath the piston during a partial brake application.

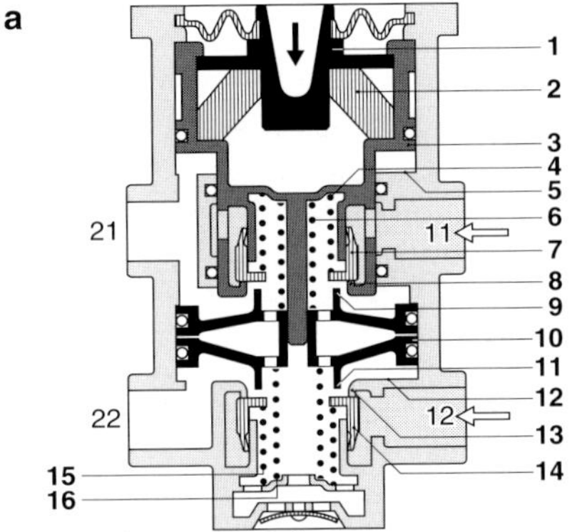

Fig. 8: Service-brake valve with rocking piston
a) Driving (non-braked) mode, b) Partially-braked mode, c) Fully-braked mode.
1 Tappet, 2 Travel-limiting spring,
3 Reaction piston, 4,15 Valve spring, 5 Stop,
6,16 Plunger spring, 7,14 Cup seal,
8,13 Inlet-valve seat, 9,11 Discharge-valve seat,
10 Rocking piston, 12 Stop.

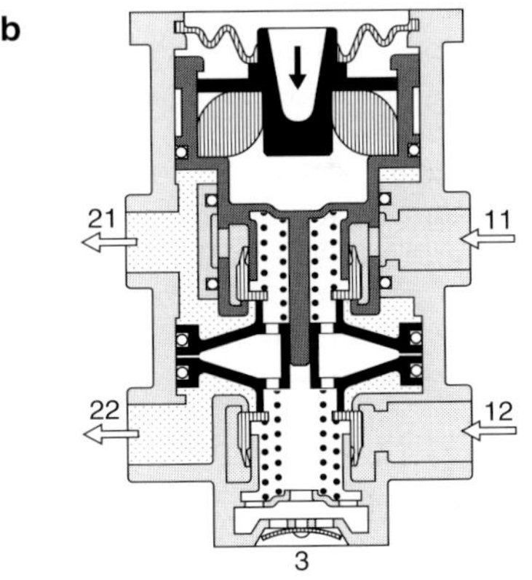

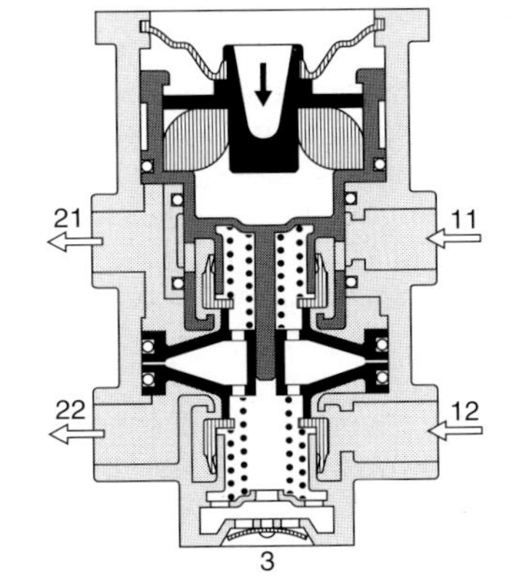

Service-brake valve with adapter valve

Yet another design – the service-brake valve with integral adapter valve – is available as an alternative to the service-brake valve with rocking piston described above.

Function

The combined service-brake and adapter valve assembly controls both of the tractor's service-brake circuits while simultaneously governing operation of the trailer's service brakes via the trailer control valve. This unit throttles the brake pressure applied to the front axle according to the pressure monitored in the rear-axle circuit.

Design

The service-brake valve with integral adapter valve shares its basic configuration with the service-brake valve with rocking piston. The difference is the integral adapter valve located between the two circuits.

This adapter valve modifies the pressure at the front axle according to the brake pressure being applied at the rear. It reduces the pressure being applied to the front brakes as a function of surface ratios.

Operation

Normal operating mode (brakes off)

During normal operation (off-state, Fig. 9), the cup seals (7, 14) remain pressed against the inlet-valve seats (8, 13). Connections 21 and 22 remain connected to the atmosphere, as does control fitting 4. The compressed air at connections 11 and 12 cannot enter the brake cylinder.

Brakes-applied mode

When the driver applies the brakes, the tappet (1) pushes against the travel-limiting spring (2) to move the reaction piston (3) downward. The discharge-valve seat (9) closes and the inlet-valve seat is pushed open, allowing air to flow from connection 11 to connection 21, which, in turn, is connected to the rear-axle brake cylinders.

The circuit between connection 21 and the brake cylinder incorporates a load-sensing valve (brake-force regulator), which reduces the pressure transmitted to the wheel-brake cylinders in accordance with the vehicle's load condition.

This brake pressure – the pressure present within the wheel-brake cylinder – is also transmitted back to connection 4 on the service-brake valve.

The tappet's movement propagates through the rocking piston (10) to close the discharge-valve seat at the cup seal (14) and push open the inlet-valve seat (13). This allows air to flow from connection 12 to connection 21.

Fig. 9: Service-brake valve with adapter valve
1 Tappet, 2 Travel-limiting spring, 3 Reaction piston, 4,16 Valve spring, 5 Stop, 6,15 Plunger spring, 7,14 Cup seal, 8,13 Inlet valve, 9,11 Discharge-valve seat, 10 Rocking piston, 12 Stop, 17 Seal ring.

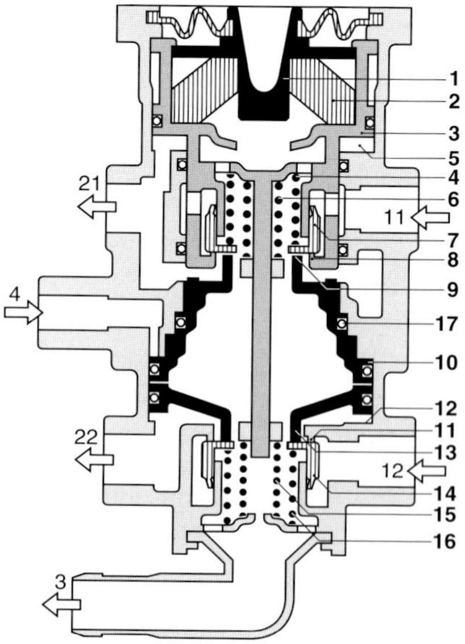

Two different pressures are exerted against the top of the rocking piston (10). (From connection 21 and connection 4; the pressure at connection 4 pushes against the annular surface between the rocking piston's outer perimeter and the pressure chamber; this pressure is limited by the seal ring [17]). A state of equilibrium is necessary between these two pressures and the pressure present at connection 22. If the vehicle is unladen, then the pressure at connection 4 will be less than pressure 21; thus pressures 21 and 4 will join to generate a combined force. This force is balanced against the force from connection 22 being exerted against the bottom of the rocking piston (10).

Fig. 10: Service-brake valve with inlet-valve seat integrated in housing

Service-brake valve with inlet-valve seat integrated in housing

Function

This service-brake valve controls two independent pneumatic brake circuits in the tractor. It also controls the trailer's service brakes via the trailer-control valve.

Design

This type of service-brake valve differs from the service-brake valve with rocking piston described above by incorporating a housing-mounted inlet-valve in brake circuit No. 1. The designs are similar in all other respects. The two circuits are arranged in series, and share a common, brake-pedal-controlled actuating mechanism. The inlet-valve seats for brake circuits 1 and 2 are installed on the valve housing.

A reaction piston installed between the two circuits serves to equalize their pressure levels when the brakes are applied. The travel-limiting spring responsible for initiating response in brake circuits 1 and 2 consists of two springs which come into effect consecutively.

Operation

Normal operating mode (brakes off)
When the service-brake valve with integral valve seat (Figures 10 and 11) is in its passive mode, the valve seats (7) and (10) remain up against the inlet-valve seats (5) and (9). The system air pressure present at connections 11 and 12 is prevented from flowing to brake connections 21 and 22, which remain connected to atmosphere via connection 3.

Brakes-applied mode
Application of pressure at the brake pedal pushes down the tappet (1). Meanwhile, the tensioned compression spring (2) exerts pressure against the reaction piston (4), moving it down until its discharge-valve seat (6) comes up against the valve seat (7). As it continues downward, the valve seat (7) lifts from the in-

actuating lever (11), which has one end connected to the vehicle axle, is pushed upward. In so doing, the cam disk (10) is rotated counterclockwise. The increasing curve radius of the cam disk pushes the roller (9) and the tappet (6), which is connected to the roller, further and further upward. With the tappet in the upper position, the input pressure at port 4 is equal to the output pressure applied to the wheel-brake cylinders. With the vehicle empty, the tappet moves to the lower position and the input pressure is applied to the wheel-brake cylinders only in a ratio of maximum 5:1. The sectional drawings depict the load-sensing valve under various vehicle loads: Fig. 13a and 13b with an empty vehicle, Fig. 13c with a fully-loaded vehicle.

If the actuating lever (11) or its linkage breaks, or the transmission between lever and axle becomes unhinged, a pre-loaded spring presses the cam disk clockwise until it comes up against the housing stop as shown in Fig. 13d. As a result, the input pressure at port 4 is applied to the wheel-brake cylinders at a ratio of 2:1 every time the brakes are applied.

<u>Principle of brake-pressure regulation</u>

During braking, compressed air flows from the service-brake valve into chamber I by way of port 4. By way of the open inlet-valve seat (14), it flows into chamber II and presses the transfer diaphragm (18) with the control plunger (3) downward. Due to this, the inlet-valve seat (5) lifts off of the valve plate (4) so that compressed air can now flow from chamber I into chamber III.

As soon as the pressure in chamber II has reached equilibrium with the force of the helical compression spring (12), the diaphragm (13) with the valve piston (1) moves upward until the pilot valve is in the center position: The inlet-valve seat (14) and the valve piston (1) are seated on the valve plate (2) so that compressed air can no longer flow from chamber I to chamber II and neither chamber is connected to the upper vent 3.

The sectional drawings show how the effective area of the fan-type piston (17) changes as a function of the vehicle load. When the vehicle is empty (Figs. 13a and 13b), the transfer diaphragm (18) makes total contact with the fan-type piston thus creating the maximum possible effective area whereas, with a half-loaded vehicle, only the inner section of the transfer diaphragm makes contact with the fan-type piston thus creating a correspondingly smaller effective area. The larger the effective area, the lower the pressure in chamber III.

When the vehicle is fully loaded (Figs. 13c), the transfer diaphragm does not make contact with the fan-type piston. However, the connection between chambers I and III is open continuously due to the upper position of the tappet (6) so that the pressure generated in chamber III is the same as that present at port 4.

The pressure in chamber III presses the relay piston (7) and the valve plate (8) downward against the force of the helical compression spring (21). This enables compressed air to flow from port 1 to the wheel-brake cylinders by way of port 2.

As soon as the pressure beneath relay piston (7) has built up to such an extent that it is in equilibrium with the pressure in chamber III, the relay piston moves upward. The valve plate (8) then closes the inlet-valve seat (20) and the discharge-valve seat (19). The relay valve is now in the center position: Compressed air can neither flow from port 1 to port 2 nor escape via the lower vent 3.

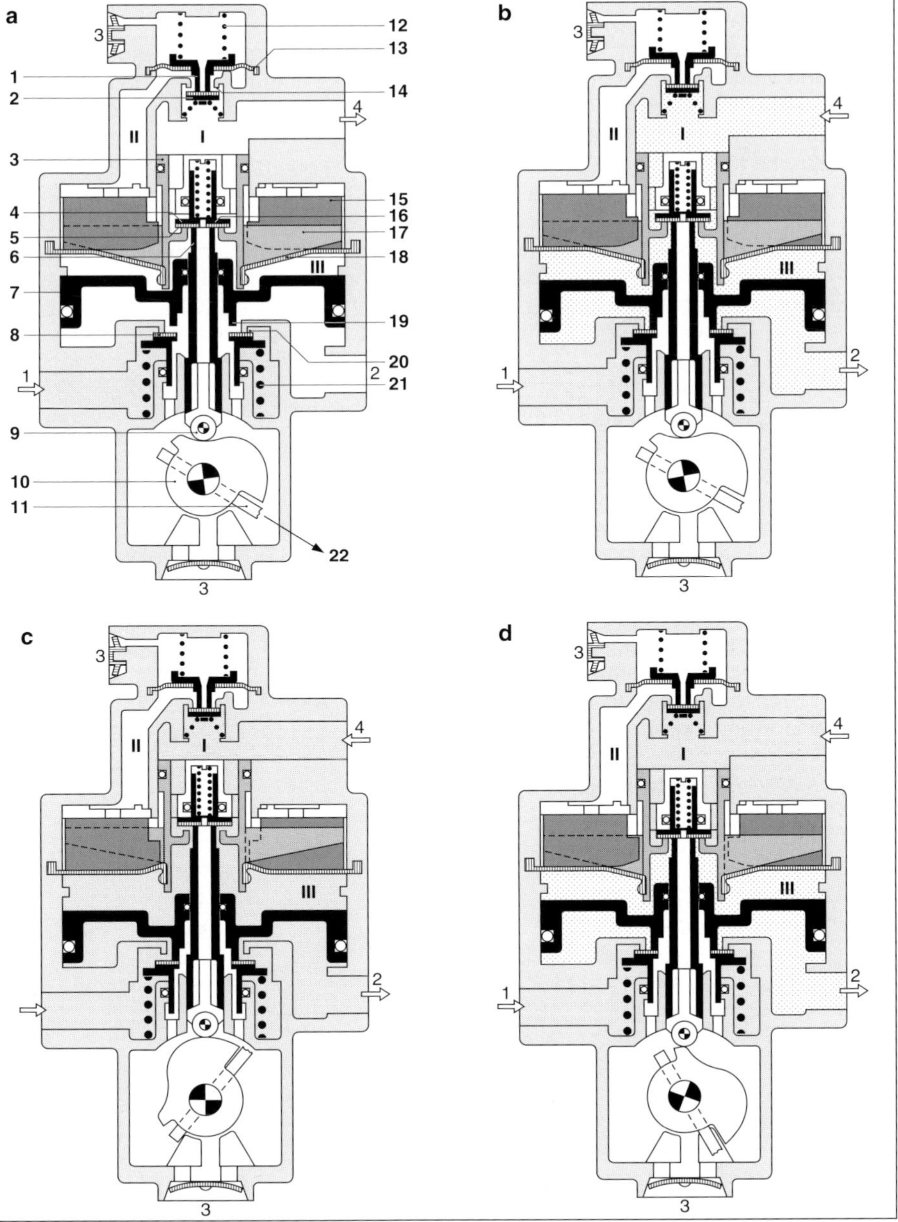

Fig. 13: Load-sensing valve (pneumatic)
a) Driving (non-braked) mode, empty vehicle,
b) Partially braked, empty vehicle,
c) Fully braked, fully loaded vehicle,
d) Fully braked, actuating lever broken.

1 Valve piston, 2,4,8 Valve plate, 3 Control plunger, 5,14,20 Inlet-valve seat, 6 Tappet, 7 Relay piston,
9 Roller, 10 Cam disk, 11 Actuating lever, 12,21 Compression spring, 13 Diaphragm, 15 Pressure plate,
16,19 Discharge-valve seat, 17 Fan-type piston, 18 Transfer diaphragm, 22 To the vehicle axle.
I ... III Air chambers.

Diaphragm actuator for wedge-actuated brakes

Purpose
This diaphragm actuator actuates the vehicle wheel brakes by converting pneumatic pressure into mechanical force.

Design
A diaphragm (6) is clamped between cylinder housing (4) and cylinder base (8). This diaphragm divides the diaphragm actuator into two chambers and applies force to the piston (5) which is firmly attached to the piston rod (3). Together with the slider (2), the piston rod is free to move in the guide tube (1). In the initial position, the force exerted by the wheel-brake wedge through the piston rod forces the diaphragm against the cylinder base.

Principle of operation
When the driver applies the service brakes, compressed air enters the diaphragm actuator through port 1 and presses against the diaphragm (6). The piston (5) extends due to the force generated by the diaphragm (brakes-applied position, Fig. 14). As a result, the piston rod (3) forces the brake wedge between the brake shoes so that they are pressed against the brake drum. When the diaphragm actuator is evacuated, the wedge immediately presses the piston rod (3) back to its initial position again (driving, non-braked, mode).

Combi-brake cylinder

Purpose
The combi-brake cylinder is a component part of the commercial vehicle's service-brake and parking-brake systems. It operates the wheel brakes for both systems every time the brakes are applied (Figs. 15 and 16).

Design
The diaphragm actuator and the spring-type brake actuator are installed one behind the other inside the cylinder housing (10). Their movement is applied to a common push rod (2) which incorporates a fork head (1) for S-cam brakes, or a slider for wedge-actuated brakes. Dirt is prevented from entering by a gaiter seal (3) fitted at the flange end of the cylinder.

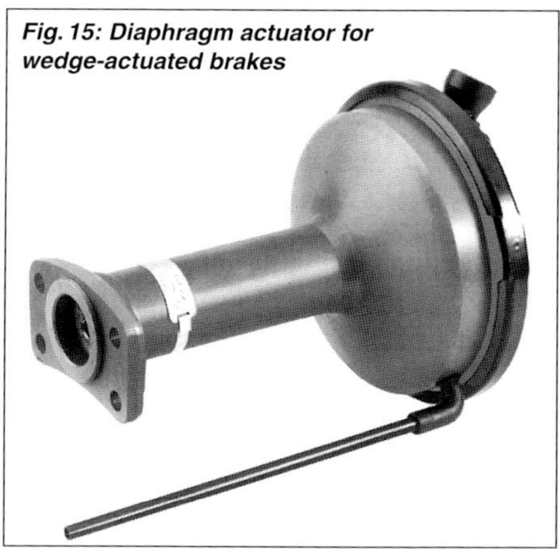

Fig. 15: Diaphragm actuator for wedge-actuated brakes

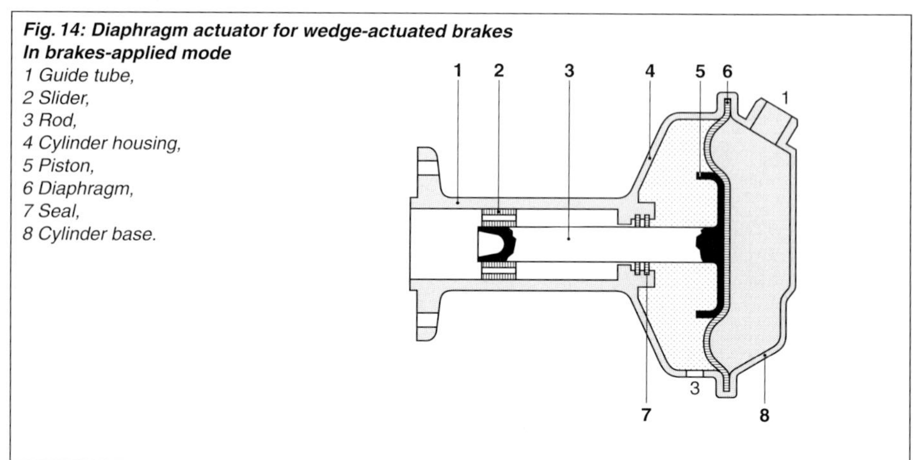

Fig. 14: Diaphragm actuator for wedge-actuated brakes
In brakes-applied mode
1 Guide tube,
2 Slider,
3 Rod,
4 Cylinder housing,
5 Piston,
6 Diaphragm,
7 Seal,
8 Cylinder base.

The spring chambers of the diaphragm-actuator and the spring-type brake actuator are connected by a channel.

Principle of operation

Driving (non-braked) mode
The diaphragm actuator is part of the service-brake system. In the driving (non-braked) mode, it forces the compression spring (4), the piston (5) with push rod (2), and the diaphragm (6), into the end position against the cylinder base. In the case of the spring-type brake actuator, which is part of the parking-brake system, in the driving (non-braked) mode, compressed air is applied to the piston (9), so that spring (8) is compressed and the spindle (7) retracts (Fig. 16a).

Brakes-applied mode
(service-brake system)
When compressed air is applied to the diaphragm (6) of the diaphragm actuator through port 11, the force of the air applied to the diaphragm shifts the piston (5) and push rod (2) against the spring force (4) and in the brakes-applied direction (Fig. 16b). The push rod initiates the braking process via the fork head (1). When the diaphragm actuator is evacuated, the compression spring (4) forces the push rod back to its initial position.

Parking-brake-applied mode
When the spring-type brake actuator is evacuated, upon being relieved of pressure the spring (8) pushes the piston (9) and the spindle (7) up against the push rod (2) which moves in the braking direction and applies the wheel brakes via the fork head (1). As soon as compressed air (at a level above the release pressure) enters the chamber in front of the piston (9) again, the piston (9) forces the spring (8) back into its initial position again (Fig. 16c). In case the compressed-air brake system is empty, or leakages have caused complete loss of compressed air, the parking-brake system actuated by the spring-type brake actuator, can be released mechanically. The spring (8) is compressed by screwing out the hexa-

gon screw (11), and this causes the push rod (2) to return to the initial position (driving (non-braked)) mode.

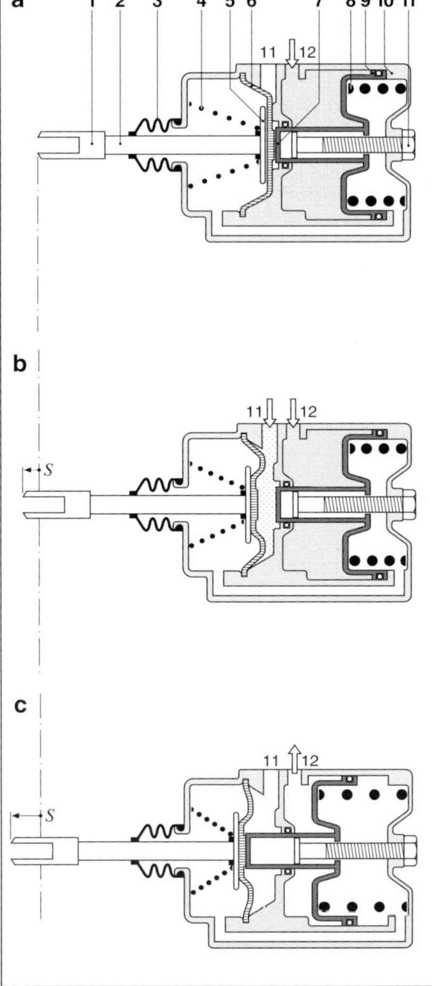

*Fig. 16: Combi-brake cylinder
(with diaphragm actuator)*
a) Driving (non-braked) mode,
b) Partially-braked mode (service-brake system),
c) Brakes-applied mode (parking-brake system).
1 Fork head,
2 Push rod,
3 Gaiter seal,
4 Compression spring (diaphragm actuator),
5 Piston (diaphragm actuator),
6 Diaphragm,
7 Spindle,
8 Compression spring (spring brake actuator),
9 Piston (spring brake actuator),
10 Cylinder housing,
11 Hexagon bolt (release device).

Dual-circuit actuator cylinder for the brake master cylinder

Function

The master cylinder dual-circuit actuator cylinder actuates a flange-mounted hydraulic master cylinder (tandem master cylinder) via mechanical linkage. The two devices convert pneumatic into hydraulic brake pressure. Hydraulic force transfer to the wheel-brake cylinders is via two fluid lines.

Design and operation

The two pistons (4, 5) are mounted consecutively within the cylinder housing (2); one of these (4) is designed to function as a drag piston. The two pistons function together to exert force against the spindle (1), which actuates the flange-mounted tandem master cylinder (Fig. 17).

Normal operating mode (brakes off)

The compression spring (3) holds both pistons (4 and 5) in their end positions.

Brakes-applied mode

When the brakes are applied, compressed air flows into the spaces behind both of the pistons, reaching one piston (5) through connection 11, and the other (drag) piston (4) via connection 12. The pressure that the service-brake valve applies at connection 11 is always greater than that entering connection 12. This pressure differential ensures that the two pistons (4,5) move outward simultaneously as they press the spindle (1) into the tandem master cylinder. Precise modulation of the force exerted by the spindle is obtained by varying the brake pressure.

Failure of a brake circuit

The force transmitted by the piston assembly remains unaffected by failure in one of the brake circuits. The drag piston (4) functions as a secondary, low-pressure unit when all circuits are intact; there is thus no change in braking performance when circuit No. 2 (connection 12) fails. If circuit No. 1 (connection 11) fails, then only the drag piston (4) will move.

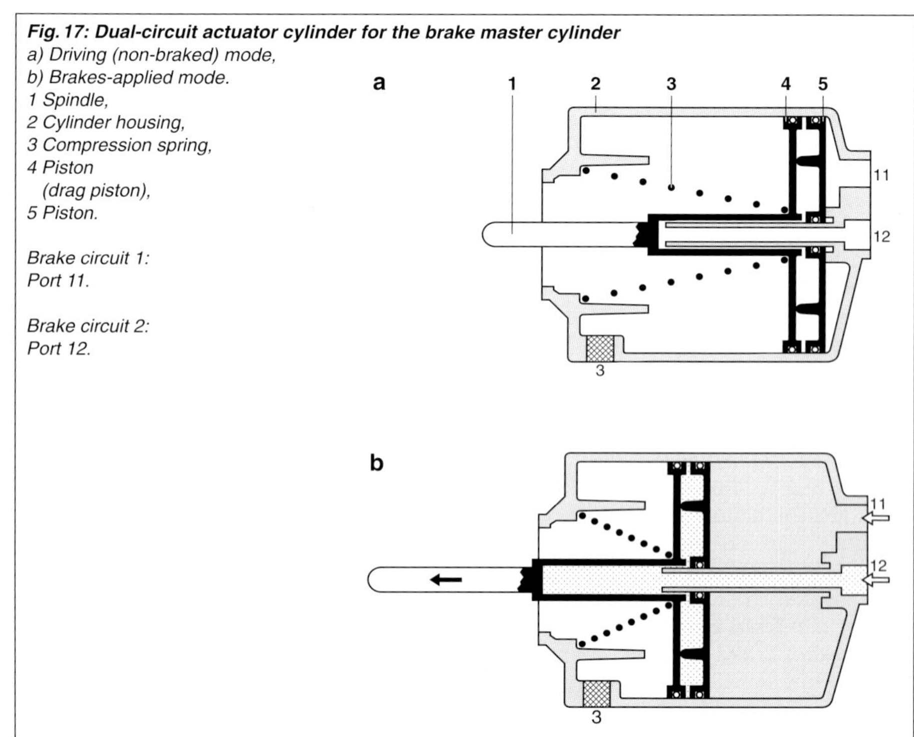

Fig. 17: Dual-circuit actuator cylinder for the brake master cylinder
a) Driving (non-braked) mode,
b) Brakes-applied mode.
1 Spindle,
2 Cylinder housing,
3 Compression spring,
4 Piston
* (drag piston),*
5 Piston.

Brake circuit 1:
Port 11.

Brake circuit 2:
Port 12.

Antilock Braking System (ABS)

Automatic load-sensitive braking-force metering (ALB) and antilock braking system (ABS)

ALB was described on the previous pages. This system controls the distribution of brake force to the individual vehicle axles as a function of the vehicle load. However, this system cannot prevent wheel lock in the event of excessive braking. This can only be achieved by the antilock braking system (ABS) which controls braking as a function of slip.

ALB and ABS cannot replace one another, but working in combination they provide optimum braking. If the vehicle is empty or only partially loaded, the ALB system reduces the brake pressure at the ALB-controlled axle and prevents excessive braking on a dry roadway. ABS intervenes when the driver applies a pressure higher than necessary for optimum braking so that the wheels have a tendency to lock.

The effect of both systems becomes particularly clear on a downgrade. Without ALB, the rear axle, which is relieved of its load to a certain degree on a downgrade, is constantly subjected to excessive braking and must therefore be continuously controlled by the ABS as a function of slip. In addition to the increased load placed on this axle's wheel brakes, more air is consumed by the constant pressurization and depressurization of the wheel-brake cylinders of the wheel pair threatening to lock.

Components of the antilock braking system (ABS)

The following components of the commercial-vehicle ABS are described below:

- Wheel-speed sensor with pulse ring
- Electronic control unit (ECU)
- Pressure control valve
- ECU for trailer recognition
- ABS plug-connection cable to the trailer.

<u>Wheel-speed sensor</u>
The wheel-speed sensing system consists of the wheel-speed sensor and the pulse ring. The pulse ring is mounted on the wheel hub and rotates at the speed of the wheel. It generates (inductively) an alternating voltage in the stationary wheel-speed sensor which is mounted on the steering knuckle or the axle-housing tube. The frequency of this alternating voltage is proportional to the rotational speed of the wheel. The pulse ring has 100 teeth, and is matched to the dimensions of the tires such as those customarily used on buses and trucks with a maximum payload exceeding 10 t as well as on trailers with compressed-air brake systems.

One pulse ring and one wheel-speed sensor are installed for each controlled wheel (Fig. 18).

<u>Electronic control unit (ECU)</u>
The ECU is the central controller. It comprises 4 functional groups.
- Input amplifier for processing the rotational-speed signals
- Computer unit for signal processing and output of the control commands
- Power stage for triggering the pressure control valves
- Monitoring circuit for checking correct operation. In the event of malfunction, the ABS is switched off, and the monitor lamp in the instrument panel lights up.

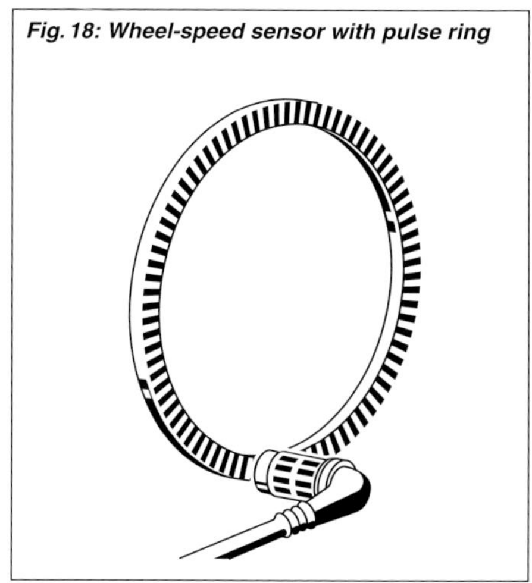

Fig. 18: Wheel-speed sensor with pulse ring

Using the signals from diagonally opposed vehicle wheels, the ECU determines a vehicle reference speed. The rotational speeds of the wheels as determined by the wheel-speed sensors are constantly compared to this reference speed. The deceleration, the acceleration, and the slip of each controlled wheel is calculated from this comparison. These three values are then used to generate the control signals for the solenoid valves of the pressure control valves which are used to provide the optimum brake pressure for the individual brake cylinders (Fig. 19).

Each vehicle is equipped with its own ECU. Since it comprises two vehicles, a tractor-trailer rig (semitrailer unit) therefore has two ECUs.

The ECU incorporates extensive error-recognition circuitry for detecting malfunctions in the entire ABS system (wheel-speed sensors, ECU, pressure control valves, wiring harness) Upon recognizing a malfunction, the ECU switches off the defective component and stores the relevant code to define the defective signal path. The service-brake system remains fully functional. When the error code is read-out in the

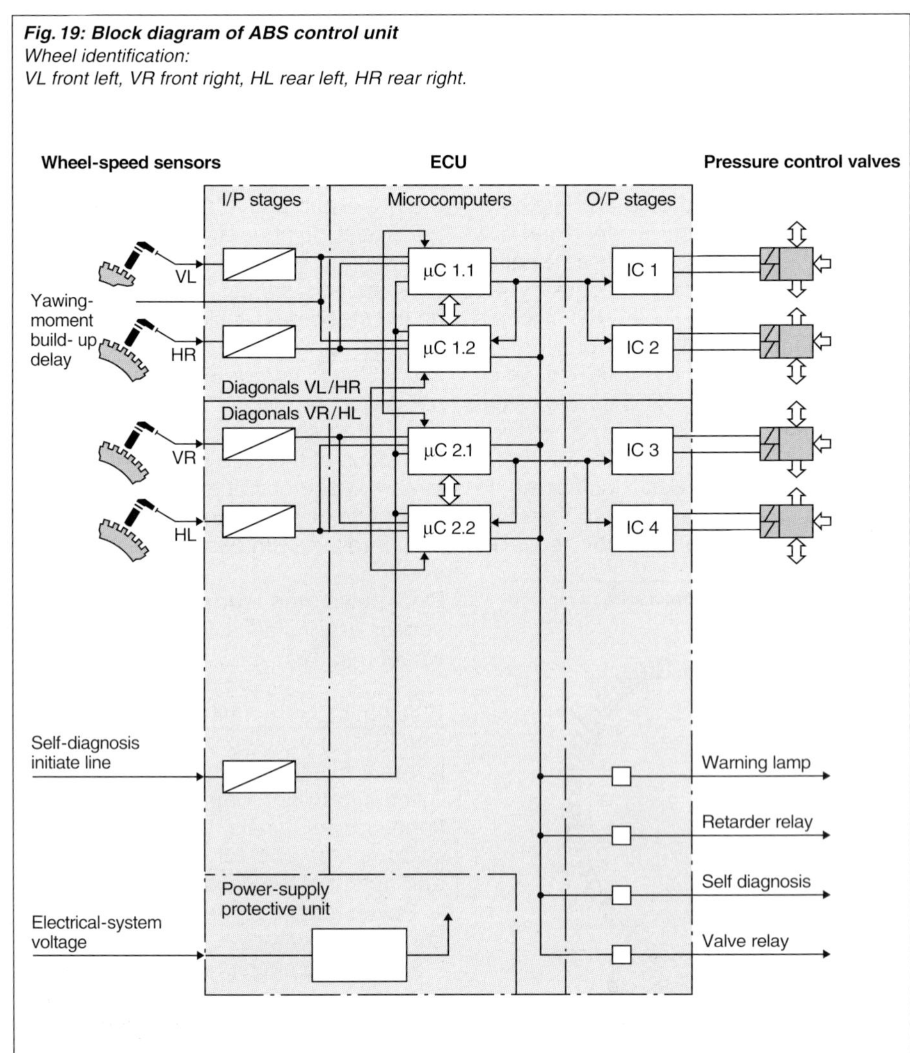

Fig. 19: Block diagram of ABS control unit
Wheel identification:
VL front left, VR front right, HL rear left, HR rear right.

workshop, the ECU's self-diagnosis (OBD) facility provides for two possibilities:

Pushbutton call-up of a display through a flashing diagnosis lamp, or call-up by means of an "intelligent" tester through a serial ISO interface.

The ABS control units incorporate not only the ABS function, but also a traction-control facility (ASR) and in some cases cruise control.

Of major importance is the fact that the ECU configures itself for the required functions. If the vehicle in question is only equipped with ABS, the ECU only performs the ABS functions. On the other hand, if the vehicle also features ASR, the ECU also performs the traction-control functions as well.

Pressure control valve

Each ABS-controlled wheel is allocated a pressure control valve (Fig. 20). This comprises two solenoid-valve controlled diaphragm valves (pressure-holding valve and outlet valve). When the brakes are applied normally, the compressed air flows to the wheel-brake cylinders through the pressure control valves which have no effect upon it at all. If, on the other hand, one of the wheels shows a tendency to lockup, the ABS control unit triggers both solenoid valves so that the pressure in this wheel's brake cylinder is reduced accordingly. The subsequent pressure-holding phase

is achieved by briefly triggering only the pressure-holding valve. Both solenoid valves are de-energized during pressure buildup.

Electronic switch unit for trailer recognition

If ABS is used in vehicle combinations, each of the two vehicles has its own ABS system. By means of the warning lamp and the reminder lamp in the tractor vehicle, the unit for trailer recognition (Fig. 21) provides the driver with the following information regarding the trailer ABS:

– Lighting of both lamps in the event of a malfunction (switch-off) of the ABS in the trailer.

– Lighting of the reminder lamp when the trailer is not equipped with ABS.

Using two warning lamps in the tractor vehicle, malfunctions in the ABS are indicated separately for the tractor vehicle and for the trailer.

ABS plug-connection to the trailer

If both vehicles of a vehicle combination are equipped with ABS, they are electrically connected to one another by means of a 5-pin ABS plug-connection cable. In a truck and trailer combination, the socket is installed in the truck and the plug connector in the trailer whereas in a semitrailer combination, the plug is in the tractor unit and the socket in the semitrailer.

Fig. 20: Pressure control valve
1 From service-brake valve, 2 To brake cylinder,
3 Pressure-holding valve, 4 Outlet valve,
5 Solenoid valve (pilot control for pressure-holding valve), 6 Solenoid valve (pilot control for outlet valve), 7 Atmosphere (3).

Fig. 21: Electronic switch unit for trailer recognition
E1 "Trailer recognition" ECU, F1 ... F3 Fuses,
H1 Info lamp, H2 "Trailer" warning lamp,
S1 Blink switch, X1 Plug connection "ABS trailer",
X2 Plug, X3 "ECU" socket.

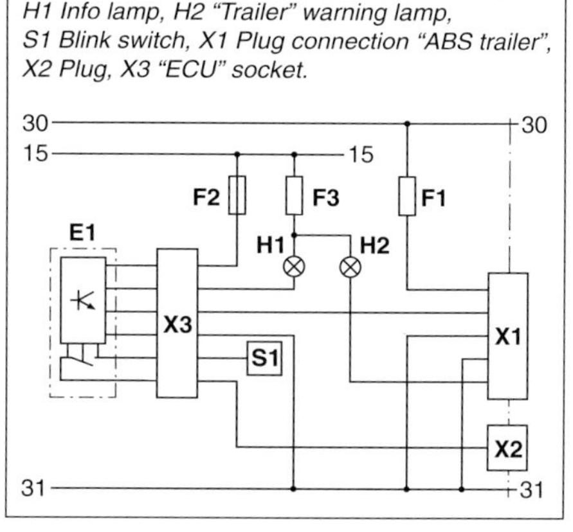

Component group C "Parking-brake system"

Parking-brake valve

The parking-brake valve controls the parking-brake system (1) in order to keep the vehicle stationary. This applies even on a downgrade, and when the driver is not present. The spring-type brake actuators are depressurized when the brakes are applied, and pressurized when the brakes are not applied

If the parking-brake system is to perform the role of a secondary braking system, a parking-brake valve which can be controlled in steps must be used. The tractor vehicle's parking-brake valve controls the trailer's service-brake system by means of the trailer-control valve.

According to EC-Guidelines, the parking-brake system thus becomes a secondary-braking system. Such secondary-braking systems must be capable of bringing the vehicle to a controlled halt with minimum brake applications of $0.25 \cdot g$.

The parking-brake valve controls the trailer's service-brake system through the trailer-control valve.

With high-pressure brake systems, the pressure of the parking-brake valve is limited to low pressure. This is because high pressure in the spring-type brake actuator is uneconomical, due to the large size of the springs which would then become necessary.

Depending upon application, two versions of the parking-brake valve are available:
– Two-position valve, and
– Three-position valve.

Using the two-position valve and the three-position valve, the spring-type brake actuators can be actuated not only in the single-circuit mode but also in the dual-circuit mode. This is carried out by a 3/2 directional control valve which is supplied with energy from a second air reservoir (e.g. secondary load).

Two-position valve

Used in single-unit trucks and buses with single-circuit actuation of the spring-type brake actuator in the following positions:
– Driving (non-braked) position (spring-type brake actuator pressurized), and
– Parking brake applied position (spring-type brake actuator depressurized)

Design

A double-seat valve is located in the housing (Fig. 2). The outlet-valve seat (5) of this valve is connected to the actuating lever (1) while the inlet-valve seat (6) is connected to the reaction piston (8).

At every position of the lever, the parking-brake valve can be adjusted in steps to apply compressed air to the spring-type brake actuators through the spring-loaded reaction piston (8).

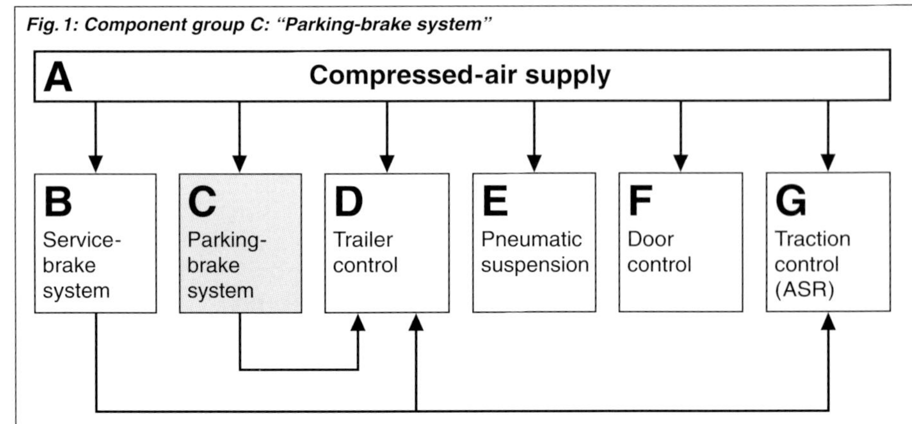

Fig. 1: Component group C: "Parking-brake system"

Operating principle

Driving (non-braked) position

When the brakes are not applied (Fig. 2), the actuating lever (1) is pulled into the left-hand end (off) position by the force of the return spring (4). The outlet-valve seat (5) makes contact with valve plate (7) so that the vent 3 is closed. The reaction spring (9) forces the inlet-valve seat (6) open against the pressure of the compressed-air applied through the reaction piston (8). The result is that port 1 is connected to port 2: The spring-type brake actuators are pressurized.

Partial-braking mode

If the actuating lever (1) is swivelled (through about 20°) to the partial braking position, the valve plate (7) follows the outlet-valve seat (5) until it makes contact with the inlet-valve seat (6). The connection between port 1 and port 2 is blocked. If the upward motion continues, the outlet-valve seat opens and connects port 2 to the vent. The outlet-valve seat remains open until the force of the reaction spring (9) exceeds the force of the air which acts from above on the reaction piston (8). The reaction piston moves upward and closes the outlet-valve seat. Inlet and outlet valves are now both closed so that equilibrium is reached, and the pressure in the spring-type brake actuator has dropped. If the actuating lever is released, the parking-brake valve returns automatically to the position used for the driving (non-braked) mode: The spring-type brake actuators are pressurized, that is, the brakes are released.

Fully braked mode

In the fully braked mode, the actuating lever (1) is pivoted until it reaches the pressure point which is easily recognizable since the required actuating force increases by approx. 100 %. In doing so, the outlet-valve seat (5) is raised by the eccentric element (3). The force of the reaction spring (9) causes the reaction piston (8) to follow this motion until the housing stop is reached, whereby it closes the inlet-valve seat (6) and opens the outlet-valve seat (5) so that the spring-type brake actuators are vented. In the process, the outlet-valve seat travels further than the reaction piston with the inlet-valve seat, and remains open as a result. The spring-type brake actuators are vented completely. When the actuating lever is released, the parking-brake valve returns automatically to the position for the driving (non-braked) mode.

Parking-brake-applied position

If the actuating lever (1) is pivoted beyond the pressure point – helical compression spring (10) is loaded – it latches into the recess of the detent element (2), and upon being released does not return to the driving (non-braked) position. The result is that the spring-type brake actuators remain vented even after the actuating lever is released.

To release the parking-brake valve, the actuating lever (1) is pulled out of the recess of the detent element (2) and swivelled back to the driving (non-braked) mode again.

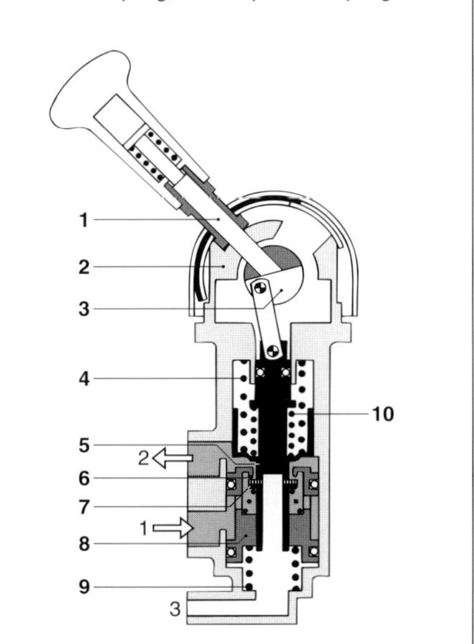

Fig. 2: Parking-brake valve (two-position valve)
Driving (non-braked) mode.
1 Actuating lever, 2 Detent element, 3 Eccentric element, 4 Return spring, 5 Outlet-valve seat, 6 Inlet-valve seat, 7 Valve plate, 8 Reaction piston, 9 Reaction spring, 10 Compression spring.

Two-position valve with 3/2 directional control valve

Used in single-unit commercial vehicles featuring two-circuit control of the spring-type brake actuators. Valve positions:
– Driving (non-braked) position (spring-type brake actuator pressurized), and
– Parking-brake-applied position (spring-type brake actuator depressurized)

Design

This component has four ports which are arranged in two planes one behind the other as shown in Fig. 3. The rear sectional plane with the ports 11 and 21 corresponds to the two-position valve which has already been described. In the front sectional plane, port 12 is connected to the secondary-load circuit while port 23 leads to the spring-type brake actuators from the 3/2 directional control valve. If the parking-brake circuit fails, the spring-type brake actuators can be vented and thus released by way of the secondary-load circuit. However, the spring-type brake actuators cannot be released in steps.

Operating principle

Driving (non-braked) position

In the driving (non-braked) mode, tappet (12) is pushed downward by cam (10) and lever (11) so that its outlet-valve seat presses against valve plate (14) and thus opens the inlet-valve seat (13). Port 12 and 23 are now connected, and the spring-type brake actuators are vented (Fig. 3).

Parking-brake-applied position

If the actuating lever (1) is moved to apply the parking brake, the force of the compressed air presses the tappet (12) and the lever (11) upward because the cam (10) also moves upward. The valve plate (14) follows this motion until it makes contact with the inlet-valve seat (13) and blocks the connection between port 12 and port 23. The tappet continues to move upward so that its outlet-valve seat opens and one port of the shuttle valve on the spring-type brake actuator is vented. The 3/2 directional control valve acts even before the pressure starts to drop at port 21.

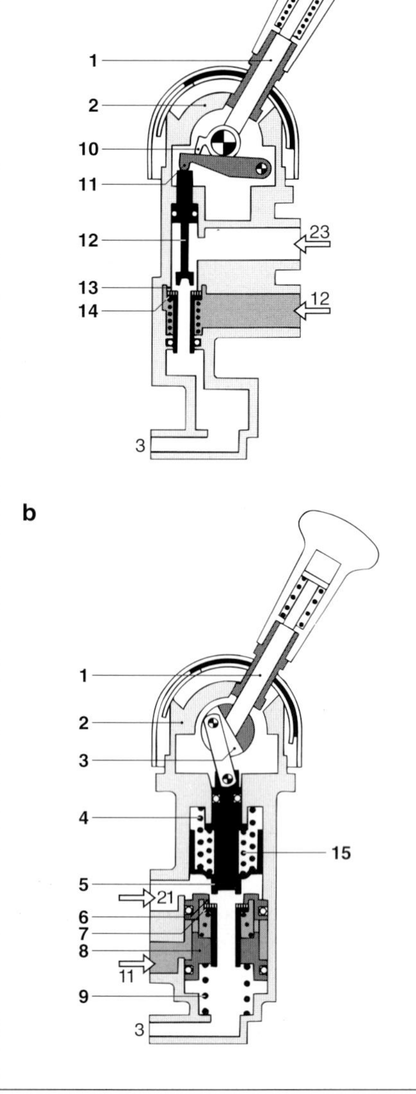

Fig. 3: Parking-brake valve (two-position valve with 3/2 directional control valve)
a) Parked mode (front sectional plane with 3/2 directional control valve),
b) Parked mode (rear sectional plane with two-position valve).
1 Actuating lever, 2 Detent element, 3 Eccentric element, 4 Return spring, 5 Outlet-valve seat, 6,13 Inlet-valve seat, 7,14 Valve plate, 8 Reaction piston, 9 Reaction spring, 10 Cam, 11 Lever, 12 Tappet, 15 Compression spring.

Three-position valve with test valve

Used in tractor vehicles with the positions:
– Driving (non-braked) position (spring-type brake actuator and control line to the trailer-control valve pressurized)
– Parking brake applied position (spring-type brake actuator and control line depressurized)
– "Test" position (spring-type brake actuator depressurized, control line pressurized).

Design

The construction of this component is identical to that of the two-position valve except that it is also provided with a test valve. Using this test valve, the trailer's service-brake system, but not the tractor's parking-brake system, can be released. In this manner, the operation of the tractor vehicle's parking-brake system can be tested on a downgrade without the trailer's service-brake system contributing to the overall braking effect.

Operating principle

From the driving (non-braked) mode up to when the parking brake is applied, the tappet (5) together with its outlet-valve seat is pressed against the valve plate (7) of the valve pipe by the cam (3) and the lever (4). The compression spring (8) keeps this outlet-valve seat closed while the inlet-valve seat (6) is open (Fig. 4a).

<u>Test position</u>

In order to actuate the test valve, the actuation lever (1) is pressed downward while in the parking-brake applied position. In this manner, the spring-loaded ratchet pin of the actuation lever emerges from the recess in the detent element (2). The actuation lever can now be pivoted even further, whereby the cam (3) releases the lever (4). The supply pressure (port 11) beneath the tappet (5) presses the tappet upward. The valve plate (7) follows this motion until it makes contact with the inlet-valve seat (6) and blocks the connection between port 21 and 22. The trailer brakes are released due to the pressurization of port 22.

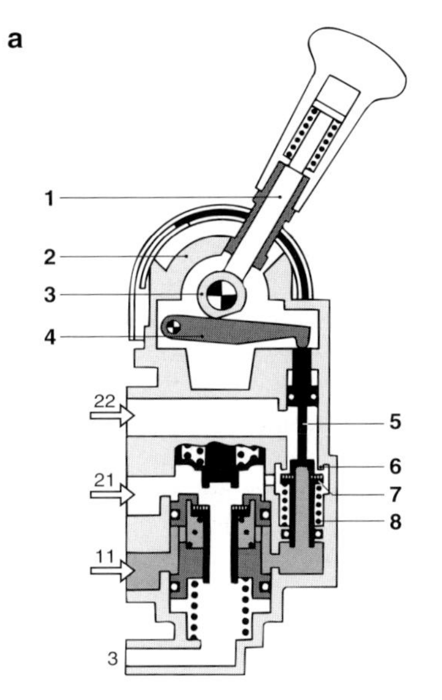

Fig. 4: Parking-brake valve (three-position valve with test valve)
a) Parked mode (tractor vehicle and trailer are braked),
b) Test position (only the tractor vehicle is braked).
1 Actuation lever, 2 Detent element, 3 Cam, 4 Lever, 5 Tappet, 6 Inlet-valve seat, 7 Valve plate, 8 Compression spring.

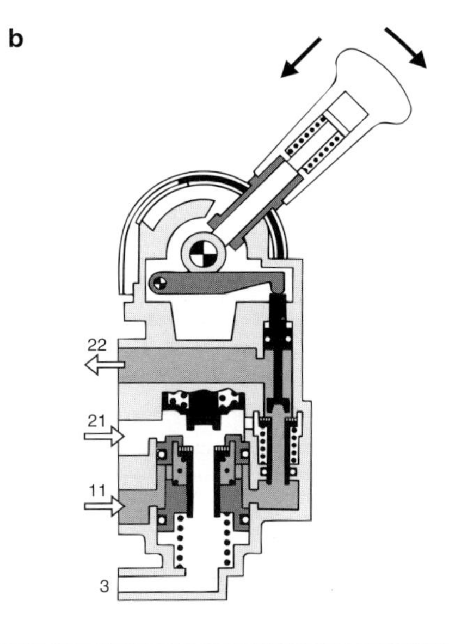

Relay valve

Purpose

The relay valve is used to pressurize and depressurize the brake cylinders at a faster rate, so that the brake response of the service-brake and/or parking-brake systems is accelerated.

Design

The relay valve contains a double seat valve which is actuated by a control line via a piston. The housing has two energy outlet ports 2 to the connected wheel-brake cylinders, one energy-supply port 1 from the air reservoir and one pilot port 4 from the brake valve (Fig. 5). A check valve or silencer in vent 3 prevents the penetration of dirt and moisture and reduces the discharge noise.

Operating principle

Small and therefore highly mobile quantities of compressed air in the low-diameter control line actuate great quantities of compressed air in the larger-diameter supply and brake cylinder lines by way of the relay valve which is located as close as possible to the wheel-brake cylinders.

When the parking-brake valve is applied, the compressed-air supply needs only to travel the short distance between the relay valve and the wheel-brake cylinders, and not the considerable distance between the parking-brake valve and the wheel-brake cylinders. In this manner, pressurization and depressurization of the wheel-brake cylinders, i.e. application and release of the brakes, is accelerated.

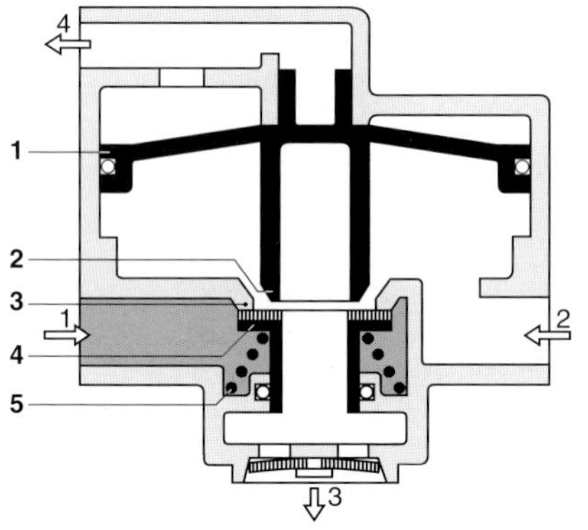

Brakes-applied mode
When not actuated, when the pilot port 4 is pressureless (pilot line depressurized, relay valve not operated), the helical compression spring (5), assisted by the supply pressure at port 1, presses the valve plate (4) against the valve seat (3) on the housing. The valve is now blocked. The line leading to the wheel-brake cylinder (port 2) is connected to atmosphere by way of vent 3 so that the wheel-brake cylinder is depressurized.

Driving (non-braked) mode (parking-brake system)
Compressed air flows through the control line (port 4) and is applied to the control plunger (1) which it presses downward. Its valve seat (2) makes contact with the valve plate (4) and blocks the connection between wheel-brake cylinder (port 2) and vent 3. The control plunger, together with its valve seat, then lifts the valve plate against the force of the helical compression spring (5) away from the valve seat on the housing (3) so that the supply air at port 1 flows into the wheel-brake cylinder line via port 2 and pressurizes the spring-type brake actuator.

In so doing, a pressure builds up beneath the control plunger. If the pilot pressure above the control plunger is equal to the pressure below it, equilibrium of the forces exists at the control plunger. The helical compression spring now presses the valve plate against the valve seat (2) again.

In this position, the valve plate (4) makes contact with the valve seats (2) and (3). The ports 1 and 2 as well as the vent are blocked with respect to one another. If the pilot pressure (port 4) drops somewhat, the equilibrium of forces at the control plunger (1) is upset.

The pressure beneath the control plun-

Fig. 5: Relay valve
1 Control plunger, 2 Valve seat on control plunger,
3 Valve seat in housing, 4 Valve plate,
5 Compression spring.

ger is now greater than the pilot pressure acting on its other side. The control plunger is pushed upward so that its valve seat (2) lifts off of the valve plate (4). As a result, compressed air flows from the wheel-brake cylinder to atmosphere by way of port 2 and vent 3 until the pressure in the wheel-brake cylinder is equal to the pilot pressure. This re-establishes the equilibrium of the forces at the control plunger. The control plunger's valve seat (2) makes contact with valve plate (4) again, and prevents any further escape of compressed air from the wheel-brake cylinder.

Spring-type brake actuator

Purpose

Spring-type brake actuators control the wheel brakes in parking-brake systems. They are particularly suited to parking-brake systems because, when depressurized, they are in the brakes-applied mode, i.e. they take effect even if the system is pressureless. They comply with legal requirements.
Spring-type brake actuators may not be used for service-brake systems.

Design

The piston (8) together with its piston rod (4) is guided in the cylinder housing (7) and is pressed to the bottom of the cylinder by the powerful outer helical compression spring (5) (Fig. 6).

Operating principle

Brakes-applied mode

As long as compressed air does not press against the piston (8) of the spring-type brake actuator, the piston rod (4) pulls the brake lever back due to the force of the powerful helical compression spring (5) which acts on the piston (8). The vehicle is braked.

Driving (non-braked) mode

To release the brakes, compressed air is applied to the piston (8) until the release pressure has been built up. The piston then moves in the direction of the stop on the cylinder housing against the force of the helical compression spring (5) and presses the brake lever toward the outside thus releasing the wheel brakes. If the compressed-air system fails, the brakes can be released by unscrewing the two hex nuts (1). As a result, bolt (2) to which the brake lever is fastened is shifted into the release position by the force of the inner helical compression spring (6) which is positioned on the piston rod (4).

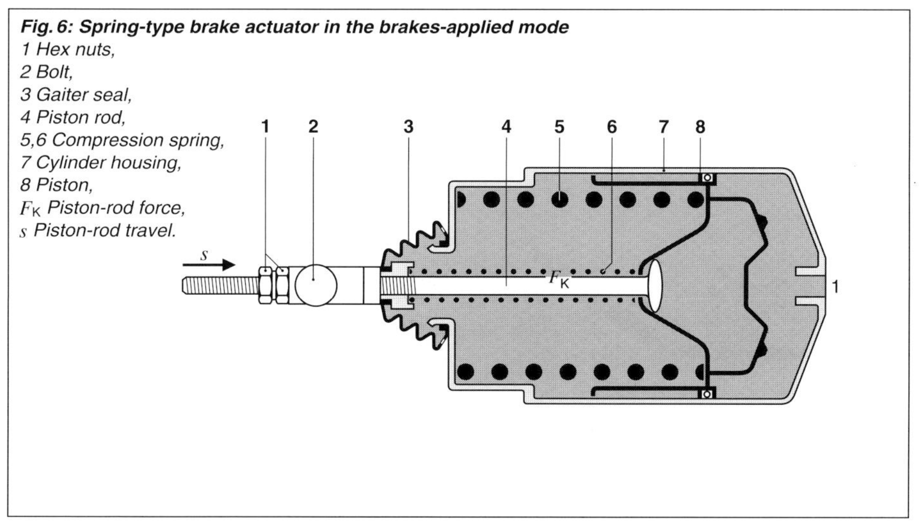

Fig. 6: Spring-type brake actuator in the brakes-applied mode
1 Hex nuts,
2 Bolt,
3 Gaiter seal,
4 Piston rod,
5,6 Compression spring,
7 Cylinder housing,
8 Piston,
F_K Piston-rod force,
s Piston-rod travel.

Component group D "Trailer control"

Trailer control valve

Purpose
In commercial vehicles with dual-line brake systems, the trailer control valve installed in the tractor vehicle controls the trailer's service-brake system and provides it with compressed air (Figs. 1, 2).

The tractor vehicle must be provided with a device which guarantees that the stipulated response time for the trailer brake system is not exceeded, should the brake line between tractor vehicle and trailer be disconnected inadvertently but not the supply line.

For vehicle combinations, the lawmakers have drawn up legislation which stipulates a given pressure tolerance in the brake line to the trailer as a function of the pressure in the tractor vehicle's service-brake system.

These requirements are fulfilled by the trailer control valve with throttle valve.

Design
The trailer-control valve for the dual-line brake system is a relay valve with three pilot chambers which actuate a double-seat valve for pressurizing and depressurizing the brake line. The two pilot cham-

bers with ports 41 and 42 (from the service-brake valve) operate with a pressure rise, and port 43 (parking-brake valve) with a pressure drop. The piston in port 41 is subdivided into two parts in order to ensure a variable pressure lead for the brake line.

The throttle valve is flanged onto the trailer control valve. The compressed air entering at port 1, flows to the "Supply" coupling head through the throttle valve and the trailer control valve (port 21)

Fig. 2: Trailer control valve

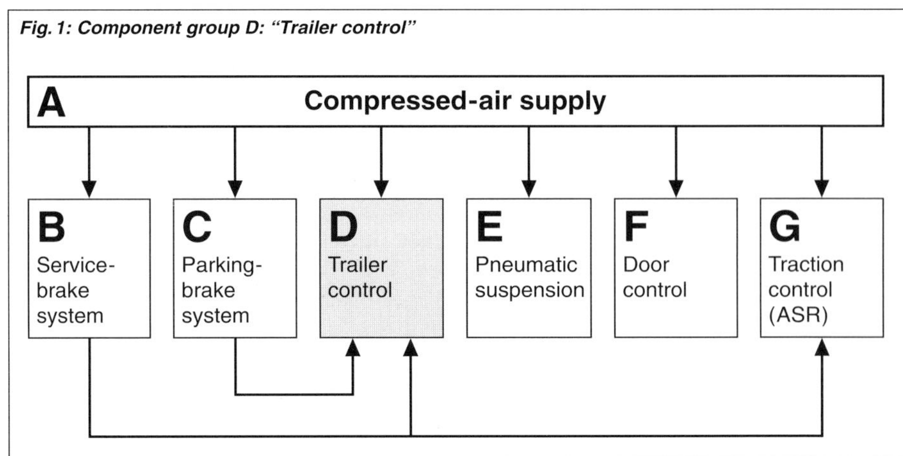

Fig. 1: Component group D: "Trailer control"

Operating principle

Driving (non-braked) mode

In the driving (non-braked mode) (Fig. 3), chambers III and IV are at the same pressure (supply pressure). Chamber III is directly connected to the air reservoir via the throttle valve, and Chamber IV to the parking-brake valve by way of pilot port 43. In view of the fact that the control plunger (18) has an effective area which is somewhat larger than that of the reaction plunger (16), the control plunger with the collar (17) is pressed against the housing stop. Assisted by the pressure of the supply air, the helical compression spring (8) holds the valve plate (15) on the inlet-valve seat (7). The control plunger (12) is held in the upper position by the helical compression spring (14). Chamber II is connected to vent 3.

Throttle valve

With no air pressure applied, the compression spring (2) forces the control plunger (3) down until the washer (6) of the (tensioned) spring assembly (4) is up against the housing stop. The force of the spring assembly (4) exceeds that of the compression spring (2), and the throttle pin (9) is in the upper position so that there is no throttling effect between port 11 and Chamber III.

Braking with the service-brake valve

During braking using the service-brake circuits, the particular control pressure is present in chambers I and V (ports 41 and 42) (Fig. 4a).

The pressure in Chamber I forces the control plunger (12) downward against the force of the compression spring (14). As a result, the outlet-valve seat (5) is closed and the inlet-valve seat (7) is pushed open. Supply air flows into the trailer's brake line 22. Along with the entry of compressed air into Chamber II, a reaction force is generated at the underside of the control plunger (12). The effective area of the control plunger (12) in Chamber I is considerably larger than that in Chamber II, and the pressures in Chambers I and II behave oppositely to their effective surfaces.

If the pressure in Chamber I is further increased, this causes an increase of pressure in Chamber II which continues until the reaction force applied to the control plunger (11) in Chamber II, exceeds the force in Chamber 1 plus the force of the spring (1). The control plunger (11) moves upward, and the control plunger (12) comes up against the housing. Since at pressures above this level only the control plunger (11) is effective, this pressure point represents the characteristic-curve knee point. The spring (1)

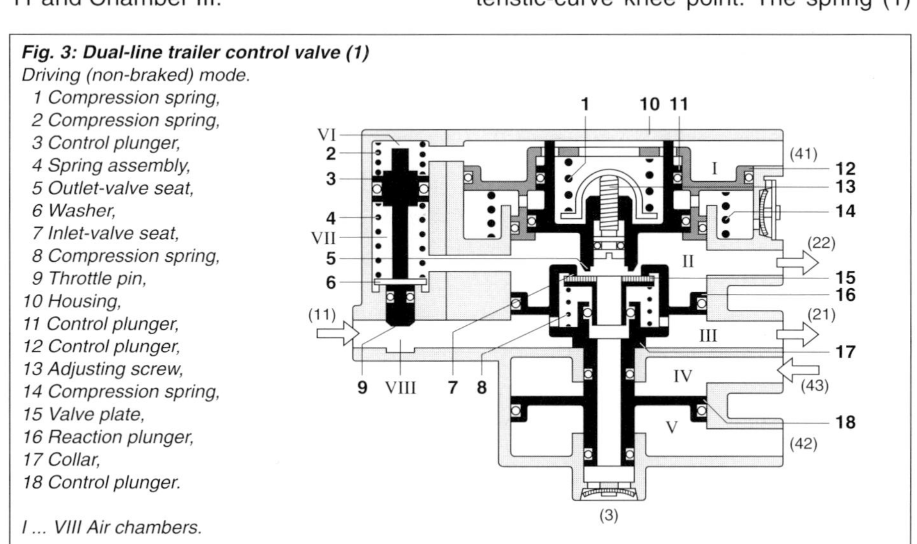

Fig. 3: Dual-line trailer control valve (1)
Driving (non-braked) mode.

  1 Compression spring,
  2 Compression spring,
  3 Control plunger,
  4 Spring assembly,
  5 Outlet-valve seat,
  6 Washer,
  7 Inlet-valve seat,
  8 Compression spring,
  9 Throttle pin,
10 Housing,
11 Control plunger,
12 Control plunger,
13 Adjusting screw,
14 Compression spring,
15 Valve plate,
16 Reaction plunger,
17 Collar,
18 Control plunger.

I ... VIII Air chambers.

pretension can be set with adjusting screw (13), so that the knee point can be changed and a higher pressure reached in the brake line.

In the partial-braking mode the downstream valve plate (15) closes the inlet valve (7). At this intermediate setting, there is no connection between the supply line (port 11), the trailer-brake line (port 22) and vent 3. In the fully braked mode, the valve plate (15) is held open by the control plunger (11) and the supply air can flow into the trailer brake line from port 11 via port 22.

If the pressure in Chamber I drops, the equilibrium of forces is disturbed and the control plunger (11) lifts. Through the open outlet-valve (5), the pressure is reduced in Chamber II and in the trailer brake line until it has reached the new control pressure.

The control pressure at port 41 is applied to the control plunger (3) via Chamber VI. Via Chamber VII, the pressure being generated in port 22 as a result of the input pressure (port 41) opposes the pressure in Chamber VI. Pressure equilibrium has been achieved and the throughflow cross-section in Chamber VIII remains unchanged (no throttling effect).

Braking with the parking-brake valve

The pressure in Chamber IV is partially, or completely, reduced when the parking brake is used for parking (Fig. 4b). Due to this pressure reduction, the force applied under the reaction plunger (16) gains the

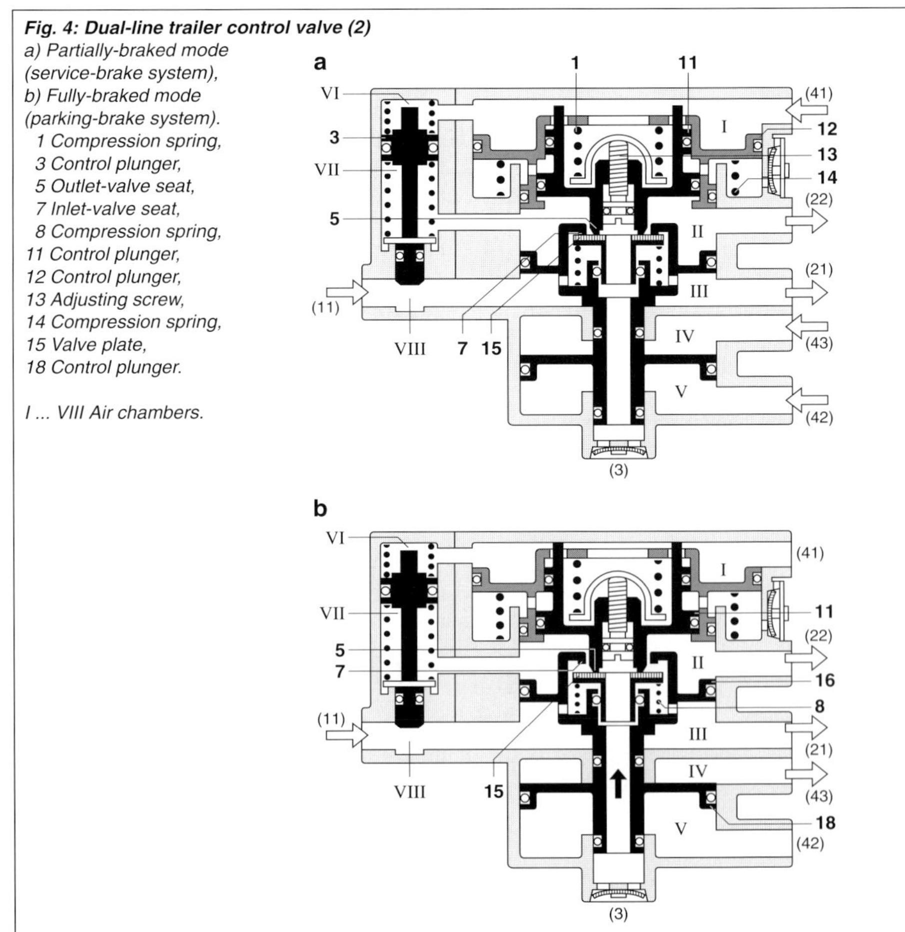

Fig. 4: Dual-line trailer control valve (2)
a) Partially-braked mode
(service-brake system),
b) Fully-braked mode
(parking-brake system).
  1 Compression spring,
  3 Control plunger,
  5 Outlet-valve seat,
  7 Inlet-valve seat,
  8 Compression spring,
11 Control plunger,
12 Control plunger,
13 Adjusting screw,
14 Compression spring,
15 Valve plate,
18 Control plunger.

I ... VIII Air chambers.

upper hand, and the plunger is forced upward. The valve plate (15) abuts against the outlet-valve seat (5) of the control plunger (11). Since the inlet-valve seat (7) moves upward with the reaction plunger (16), supply air from Chamber III can enter Chamber II through the gap between valve plate (15) and inlet-valve seat (7), from where it flows to port 22 and the trailer-brake line. At the same time, the trailer-brake pressure being generated in Chamber II is applied to the top side of the reaction plunger (16).

During partial braking, as soon as the pressure rise in Chamber II corresponds to the pressure drop in Chamber IV, the reaction plunger (16) with the inlet-valve seat (7) is pressed against the valve plate (15) without lifting the valve plate off of the outlet-valve seat (5). Because of this, the inlet and outlet are now closed and the valve plate is in the center position.

If the braking action is to be reduced, the parking-brake valve regulates a higher pilot pressure by way of Port 43 into Chamber IV of the trailer-control valve. The control plunger (18) which presses downward, lifts the valve plate (15) off of the outlet-valve seat (5) until the trailer braking pressure in chamber II has dropped by the same ratio as the pilot pressure in chamber IV has increased. The reaction plunger (16) is then pushed upward by the pressure difference between chambers III and II until the valve plate is in the center position. The inlet and outlet are closed.

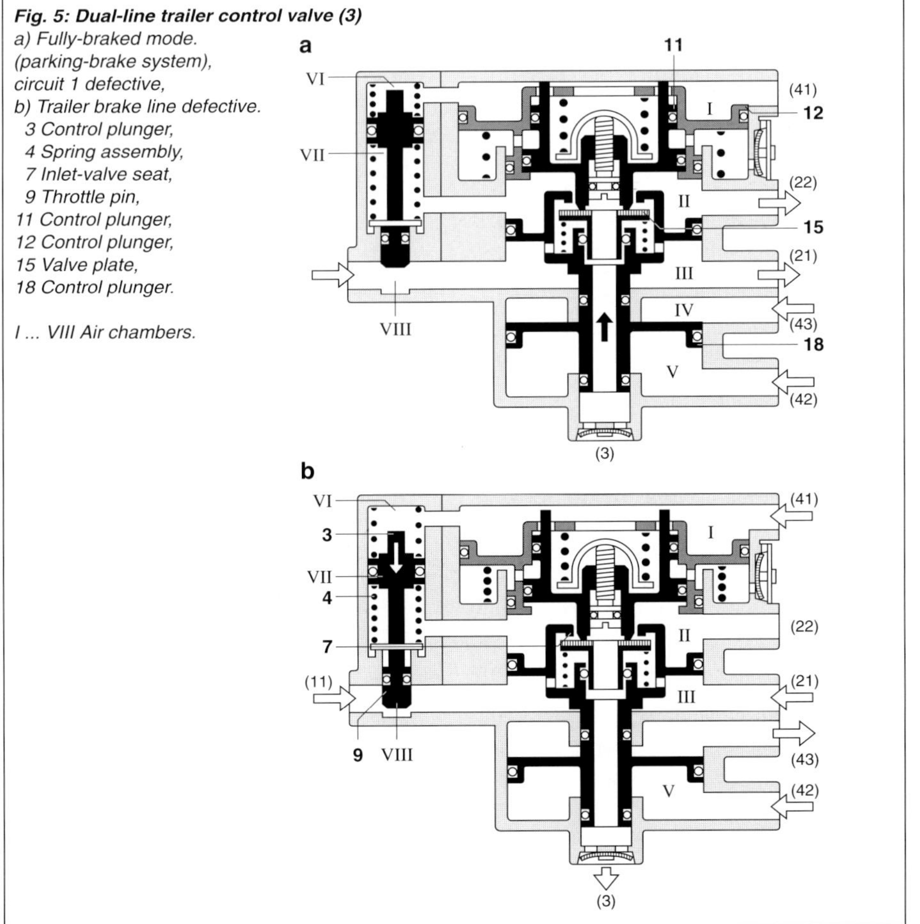

Fig. 5: Dual-line trailer control valve (3)
a) Fully-braked mode.
(parking-brake system),
circuit 1 defective,
b) Trailer brake line defective.
3 Control plunger,
4 Spring assembly,
7 Inlet-valve seat,
9 Throttle pin,
11 Control plunger,
12 Control plunger,
15 Valve plate,
18 Control plunger.

I ... VIII Air chambers.

Defective service-brake system

<u>Brake circuit 1 defective</u>
In the event of a failure of brake circuit 1, the control plunger (18) assumes the control function. This is in contrast to an intact system where this control plunger does not perform a switching motion when braking with the service-brake valve (Fig. 5a).
Control plunger (18) has a smaller effective area than control plunger (12). In addition, it must overcome greater frictional forces. This means that in the event of a failure of circuit 1, the pilot pressure necessary to provide the same trailer braking pressure is higher than required with an intact system.

<u>Brake circuit 2 defective</u>
If brake circuit 2 (port 42) fails, the trailer-control valve operates in the same manner as with an intact brake system.

<u>Trailer-brake line
(port 22) defective</u>
When the service brakes are applied, a pressure builds up at port 41 which is applied to the control plunger (3) via Chamber VI (Fig. 5b). In case of fracture of the brake line to the trailer, no pressure can build up at port 22 and in Chambers II and VII. This means that the air pressure in Chamber VI exceeds the force of the spring assembly (4). This causes the control plunger (3) to move downward, and the throttle pin (9) enters Chamber VIII until it comes up against the housing. The throttle pin (9) inside Chamber VIII reduces the cross section between ports 11 and 21. At the same time, since pressure cannot build up in Chamber II due to the fracture in the brake line, the inlet valve remains open and supply air escapes from the trailer via port 21, inlet-valve seat (7), and vent 3. As a result, the pressure in the supply line drops rapidly so that the trailer brake valve automatically triggers trailer braking, as is the case when the trailer inadvertently becomes disconnected.

Coupling heads "Supply" and "Brakes"

Design of the "Supply" coupling head

<u>Tractor-vehicle coupling head (supply)</u>
The supply coupling head has a line port from the trailer-control valve's supply line, and a coupling port for the supply line to the trailer which is protected by means of the swivel cover (2).
The stop lug (4) on the claw (5) prevents the interchange of the supply and brake lines during coupling. A shutoff valve (6 and 7) prevents supply air from escaping when the trailer is uncoupled (Fig. 7a).

<u>Trailer coupling head (supply)</u>
The coupling head has a threaded port for the trailer supply line, and a coupling port for the supply line to the tractor vehicle. It does not have the shutoff element or the swivel cover. The stop lug (10) in the claw (9) prevents interchanging the trailer lines during coupling.

Important:
Coupling-head color code:
"Supply" = Red (ISO 1728)

Fig. 6: Coupling head "Supply"

Operating principle of the "Supply" coupling head

When uncoupled, the valve plate (7) of the tractor-vehicle coupling head is pressed against the valve seat (6) by the helical compression spring (8), assisted by the supply pressure. The supply line is blocked.

During coupling of the trailer coupling head, segment-shaped claws (5, 9) engage in the claw guide (1, 12) of the tractor-vehicle coupling head. The two coupling heads are locked by rotating the trailer coupling head to the stop. The seal ring (11) of the trailer coupling head is thus firmly pressed against the thrust member (3) of the tractor-vehicle coupling head and seals off the two heads with respect to one another. At the same time, the thrust member is pressed downward and forces open the valve plate (7) so that supply air flows to the trailer.

Design of the "Brakes" coupling head

Tractor-vehicle coupling head (brakes)
The brakes coupling head is identical to the supply coupling head. The brake-line stop lug prevents interchange of the brake and supply lines during coupling.

Trailer coupling head (brakes)
The brakes coupling head has one line port for the trailer brake line, and a coupling port for the brake line to the tractor vehicle. It is not provided with a shutoff valve but it does have a swivel cover. The stop lug prevents the unintentional connection of the trailer brake line to the tractor-vehicle supply line when coupling the trailer.

Important:
Coupling-head color code:
"Brakes" = Yellow (ISO 1728)

Operating principle of the "Brakes" coupling head

The operating principles of the "brakes" and "supply" coupling heads are identical.

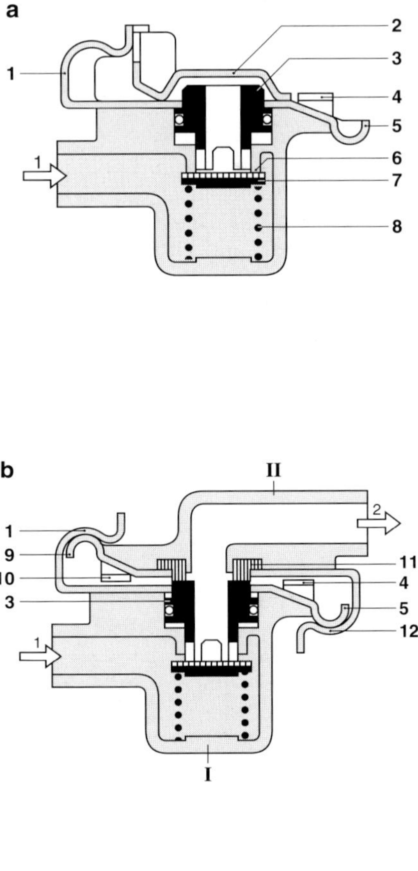

Fig. 7 Coupling head "Supply/Brakes"
a) Tractor-vehicle coupling head
"Supply" or "Brakes"
Shutoff valve closed.
b) "Supply" or "Brakes" coupling heads connected between tractor vehicle and trailer.
Shutoff valve opened.
1 Claw guide, 2 Swivel cover,
3 Thrust member,
4 Stop lug, 5 Claw, 6 Valve seat,
7 Valve plate, 8 Compression spring,
9 Claw, 10 Stop lug,
11 Seal ring, 12 Claw guide.
I Tractor-vehicle coupling head,
II Trailer coupling head.

Component group E "Pneumatic suspension"

Purpose

To cushion the vehicle body on an uneven roadway and to maintain a constant clearance from the roadway under all load conditions.

Height-control valve with damping

Purpose

The height-control valves regulate the quantity of air in the air-spring bellows in accordance with the vehicle load so that the clearance between vehicle body and axle, or between body and road surface, always remains constant.

Design

The height-control valve is mounted on the vehicle body and connected to the vehicle axle by an actuating lever and a connecting rod. The valve consists of three groups which act in parallel to one another.

– Actuating device
– Control valves
– Delay device (damping)

Operating principle
The operating principle is explained by referring to Figs. 2 and 3.

Basic principle
The actuating device acts on the control valves when the vehicle body moves upward or downward due to a change in vehicle load. By pressurizing or depressurizing the air-spring bellows, the control valves maintain constant clearance between vehicle body and road surface. The delay device slows down the control-valve response, thus providing a damping effect when unevenness in the road surface causes brief motions of the actuating lever (5).

Valve in center position
With the actuating lever (5) in the horizontal center position; the height-control valve is also in the center position, i.e. the inlet valve (8) and the outlet valve (6) are closed.

Vehicle load is increased
(air-spring bellows are pressurized)
The actuating lever (5) is rotated upward by the connecting rod which is connected to the vehicle axle. At first, the actuating shaft (1) does not follow the movement

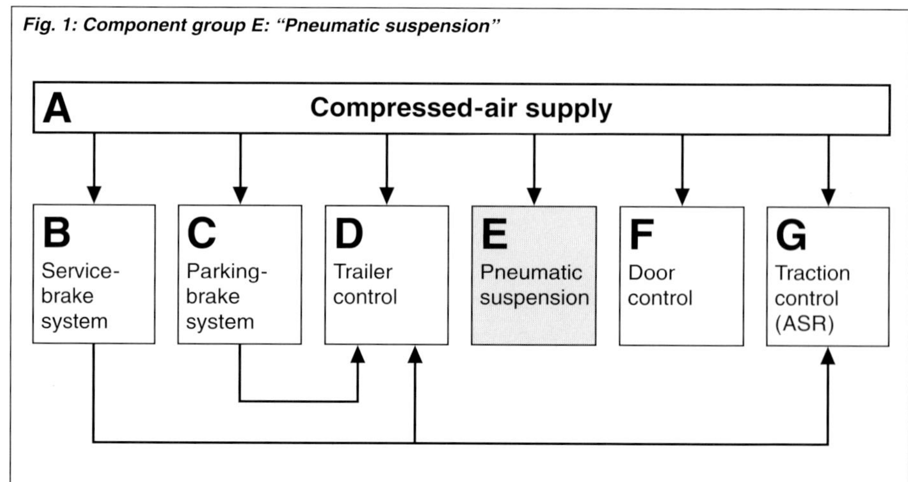

Fig. 1: Component group E: "Pneumatic suspension"

of the actuating lever due to the delay device. The spring piston (3) which is pre-loaded by the helical compression spring (4) is lifted, presses against the edge of the flattened actuating shaft and deflects it, whereby the spring force increases along with the degree of deflection. Thus, a torque acts on the actuating shaft and thus also on the delay device. The actuating shaft (1) attempts to drive the piston (12) by way of the driver pin (9). The piston resists this motion, however, because the damping chambers beneath the piston are filled with fluid. When moving, the piston must displace this fluid through the small gap between piston and housing from one damping chamber to the other. If the piston (12) is pressed in the opening direction of the inlet valves (8), the corresponding rotational motion of the actuating shaft (1) is transmitted to the inlet valve (8) by way of the control pin (7). Once the control pin has travelled a certain distance, it pushes open the inlet valve through which compressed air can now flow into the air-spring bellows. This results in the lifting of the vehicle body.

As the vehicle body lifts, the actuating lever (5) resumes its horizontal center position again, thus causing the inlet valve (8) to close again, i.e. no more compressed air can flow to the air-spring bellows.

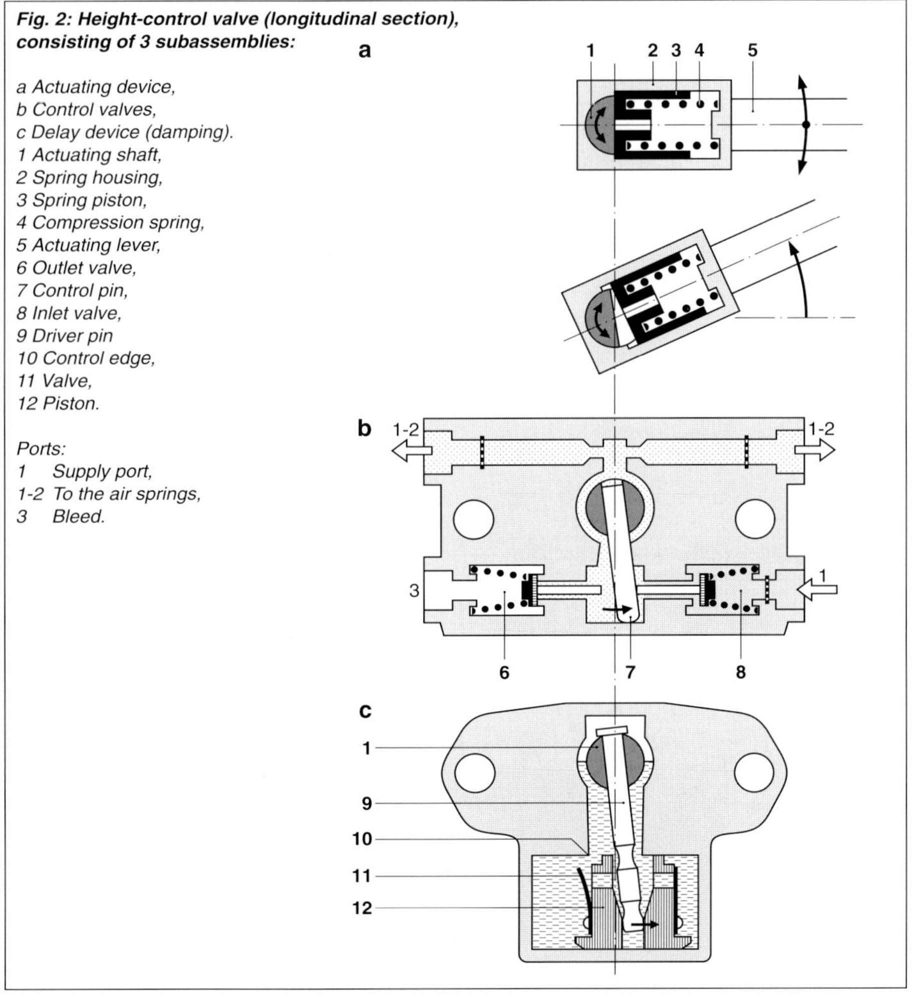

Fig. 2: Height-control valve (longitudinal section), consisting of 3 subassemblies:

a Actuating device,
b Control valves,
c Delay device (damping).
1 Actuating shaft,
2 Spring housing,
3 Spring piston,
4 Compression spring,
5 Actuating lever,
6 Outlet valve,
7 Control pin,
8 Inlet valve,
9 Driver pin
10 Control edge,
11 Valve,
12 Piston.

Ports:
1   Supply port,
1-2 To the air springs,
3   Bleed.

The vehicle body has achieved its original clearance to the vehicle axle or to the roadway again.

In the delay device, the driver pin (9) can press the piston (12) out of its off position and into the center position again without any great piston resistance. This is because the fluid can rapidly flow back through the large gap between the control edge (10) and the piston as well through the valve (11) which opens. The piston follows the speed of the actuating lever (5).

Vehicle load is decreased
(air-spring bellows are depressurized)
As the vehicle body lifts when the vehicle is being unloaded, the actuating lever (5) rotates downward and opens the outlet valve (6) after overcoming the damping resistance. Compressed air now flows out of the air-spring bellows to atmosphere through the outlet valve (6) until the height-control valve returns to its center position.

Air-spring bellows
(U-bellows)

Purpose
The air-spring bellows mounted on the vehicle axle carry the vehicle body, and due to their elastic response function as suspension elements. In view of the fact that the air pressure present in the air-spring bellows is regulated in accordance with the vehicle load, the air-spring bellows provide optimum damping of the vehicle body and constant clearance between it and the road surface under all vehicle load conditions.

Design
The cylindrical U-bellows are made of rubber with a woven-fabric insert. This fabric consists of synthetic fibers which are arranged as support members in the form of a horizontal cross. The steel piston which closes off the bellows from below is attached to the vehicle axle by means of a bracket. A plate above the bellows holds the vehicle body and contains ports for the supply and outlet of compressed air.

Operating principle
The vehicle body is borne by the compressed air in the bellows and in the piston. The height-control valve supplies compressed air as the vehicle load increases and removes it as the load decreases.

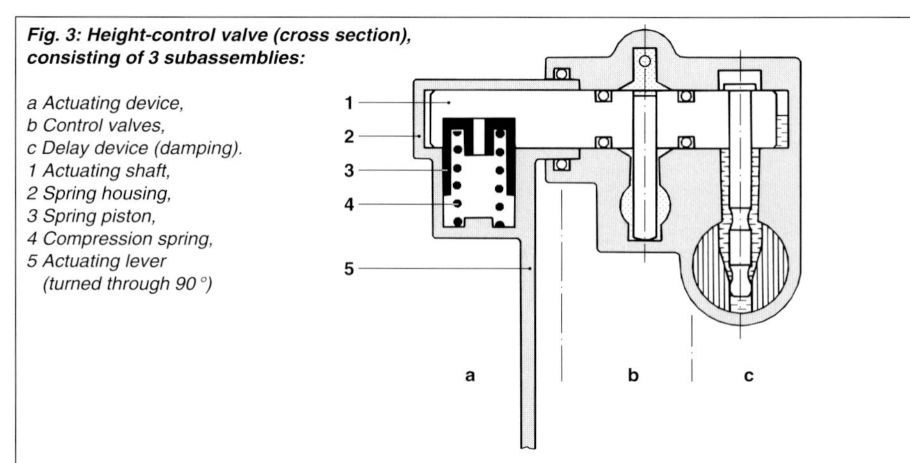

*Fig. 3: Height-control valve (cross section),
consisting of 3 subassemblies:*

*a Actuating device,
b Control valves,
c Delay device (damping).
1 Actuating shaft,
2 Spring housing,
3 Spring piston,
4 Compression spring,
5 Actuating lever
   (turned through 90 °)*

Electronically-controlled pneumatic suspension (ELF)

Function
The pneumatic suspension system with electronic control (Elektronisch geregelte Luftfederung, or ELF) monitors and maintains vehicle height as a function of the distance between body and axles. The system controls the solenoid valves in a directional-control valve assembly in order to maintain the vehicle at a defined mean height under all load conditions.

Design
The electronically-controlled pneumatic suspension system (Figures 4 and 5) incorporates the following three main components:

– level sensor,
– electronic control unit (ECU), and
– directional-control valve assembly with solenoid valves (also available with "kneeling" function to lower buses on the entry/exit side).

Operation

Level sensor
The level sensor employs a rotary potentiometer, featuring a contact wiper (brush) connected to the vehicle axle via transfer rod, to monitor the distance between vehicle axle and body.

Electronic control unit
The plug-in interface is responsible for relaying the voltage signals from the level sensor to the electronic control unit.
The air suspension elements compress more and more as the vehicle is loaded. The transfer rod from the vehicle axle rotates the wiper along the potentiometer track to the "raise" position. The control unit recognizes a voltage deviation and responds by transmitting position-control commands to the appropriate solenoid valves. The solenoid valves then deliver compressed air into the suspension elements to raise the vehicle. The corresponding principle applies as the vehicle

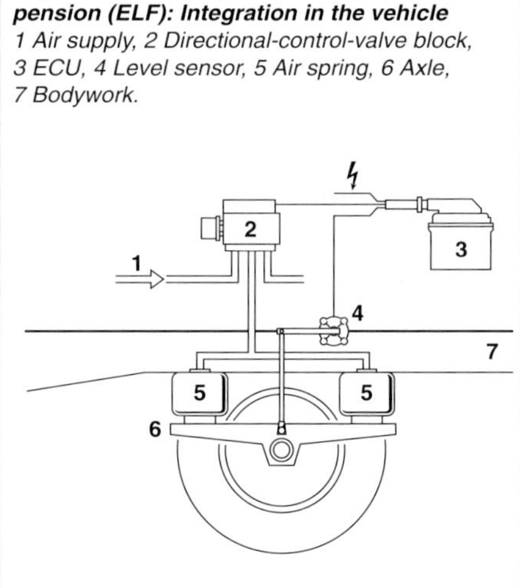

Fig. 5: Electronically-controlled pneumatic suspension (ELF): Integration in the vehicle
1 Air supply, 2 Directional-control-valve block, 3 ECU, 4 Level sensor, 5 Air spring, 6 Axle, 7 Bodywork.

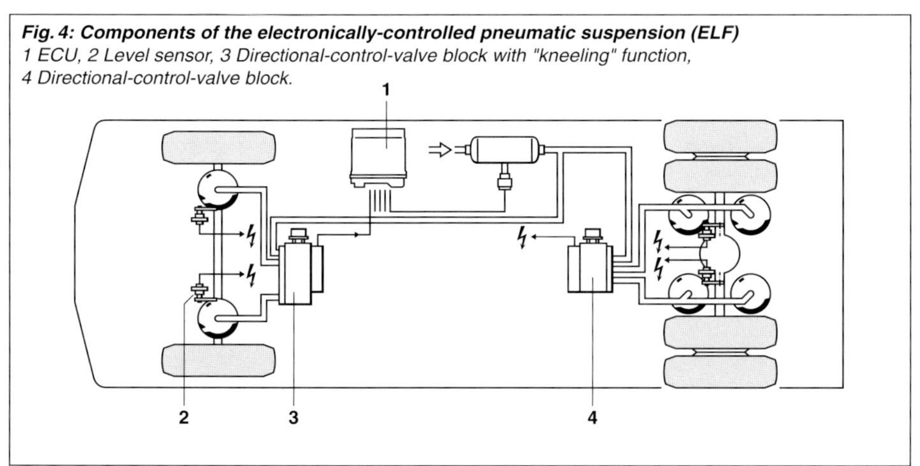

Fig. 4: Components of the electronically-controlled pneumatic suspension (ELF)
1 ECU, 2 Level sensor, 3 Directional-control-valve block with "kneeling" function, 4 Directional-control-valve block.

is unloaded. Air flows into or out of the suspension bellows until the level sensor indicates that the vehicle body has returned to its standard ride height.

Triple directional-control-valve block

The air flow required to raise or lower the vehicle bodywork is governed by a triple directional-control-valve assembly (Fig. 6). This control-valve block generally includes three solenoid valves mounted at the rear axle.

The central solenoid valve, a 3/2 directional-control unit (A), controls the flow of air into and out of the air suspension elements, while the downstream-mounted 2/2 solenoid valves (B, C) are assigned to one control circuit per vehicle side.

None of the solenoids is energized in the basic position for the normal operating mode, in which the solenoid armatures (1, 2, 3) remain up against the inlet-valve seats (4, 5, 6). There is no pressure in chambers I, II and III above the relay pistons (7, 8, 9), while chamber IV is also depressurized via connection 3 (atmosphere). The valve plates (10, 11, 12) are up against the inlet-valve seats (13, 14, 15); connections 1, 21 and 22 are thus closed.

While the vehicle is being loaded, the level sensors move to the "raise" position. The control unit responds by activating the solenoid valves (A), (B) and (C), lifting the solenoid armatures (1, 2, 3) from their respective inlet-valve seats (4, 5, 6). This insulates chambers I, II and III from the discharge circuit, allowing compressed air to enter.

The relay pistons (7, 8, 9) move downward. Initially, the discharge-valve seat (16) is pressed against the valve plate (10) at the center solenoid valve (A), cutting off the connection with the outside atmosphere. The relay piston (7) opens the inlet-valve seat (13) and compressed air flows into chamber IV. Since relay pistons (8) and (9) are also activated, the descending relay pistons lift the valve plates (11) and (12) from their inlet-valve seats (14 and 15). A stream of compressed air enters the air suspension elements through connections 21 and 22, continuing until the vehicle returns to its standard ride height. The current to solenoids (A), (B) and (C) is then interrupted and the valves in the directional-control block revert to their normal operating-mode position.

While the vehicle is being unloaded, the level sensors move to the "discharge" position. The control unit activates the

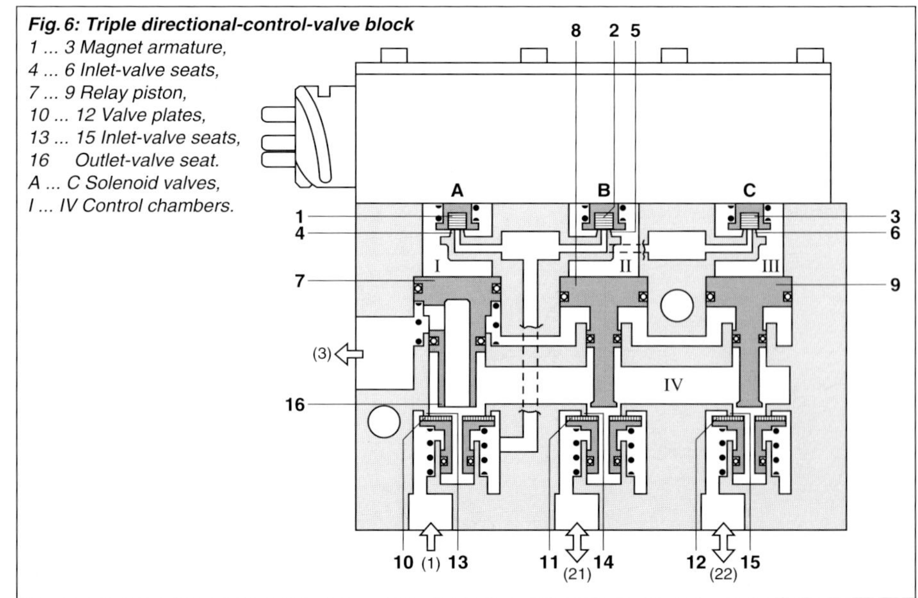

Fig. 6: Triple directional-control-valve block
1 ... 3 Magnet armature,
4 ... 6 Inlet-valve seats,
7 ... 9 Relay piston,
10 ... 12 Valve plates,
13 ... 15 Inlet-valve seats,
16   Outlet-valve seat.
A ... C Solenoid valves,
I ... IV Control chambers.

solenoids (B) and (C). The relay pistons (8) and (9) open the inlet-valve seats (11) and (12). Compressed air flows from connections 21 and 22 and through connection 3 before escaping to atmosphere. This process continues until the vehicle resumes its standard ride height, at which point the current to the solenoids (B) and (C) is interrupted.

Triple directional-control-valve block with "kneeling" function and cross throttle

The block assembly incorporating the triple directional-control valves with "kneeling" function (Fig. 7) is generally installed on the front axle, and includes three solenoid valves to control the following functions:

– Solenoid valve (A) is the central solenoid valve, and regulates the supply and discharge of air to and from the air suspension elements.

– Solenoid valve (B) controls the flow of air to connections 21 and 22. In their deactivated state these two connections are joined by a cross throttle (6).

– Solenoid valve (C) blocks the air flow to connection 22 as required.

In operation, the differences between this unit and the "standard" directional-control-valve block described above are as follows:

In the basic position for normal operation, a cross throttle provides a link between connections 21 and 22.

While the vehicle is being unloaded, the electronic control unit activates the solenoid valve (B). Relay pistons (1) and (2) lift the valve plates (4) and (7) to open a passage between the two connections 21 and 22.

Solenoid valve (C) is not activated during height adjustment.

The "kneeling" mode is intended to facilitate passenger entry and exit by lowering the access-side of the bus body. To start this mode, solenoid valves (B) and (C) are activated. Solenoid valve (C) activates the relay piston (3) to close the passage leading from connection 21 to connection 22. Meanwhile, solenoid valve (B), which was energized at the same time, activates its relay piston (1) to open the inlet-valve seats (5) and (8). Compressed air flows through connections 21 and 3 on its way to atmosphere. The bus body continues to sink until it reaches kneeling height. The current supply to solenoid valve (B) is then deactivated and the kneeling process is terminated.

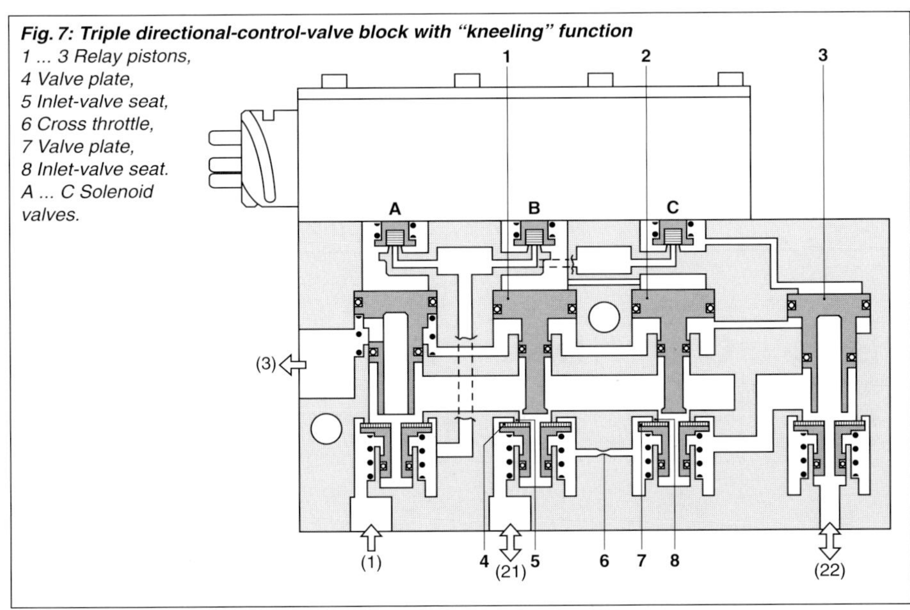

Fig. 7: Triple directional-control-valve block with "kneeling" function

1 ... 3 Relay pistons,
4 Valve plate,
5 Inlet-valve seat,
6 Cross throttle,
7 Valve plate,
8 Inlet-valve seat.
A ... C Solenoid valves.

Component group F "Door control"

Purpose

To open and close the vehicle's doors.

Solenoid valve

Purpose

The solenoid valve converts electrical control pulses into pneumatic control pulses. In the door-control system it permits remote control of the double-acting working cylinders for opening and closing the vehicle doors. It alternately supplies compressed air into one cylinder chamber while depressurizing the other chamber.

Design

The main components of the solenoid valve are the electromagnet (1), the rocker (3) and the inlet and outlet valves (4 and 5) which are connected with one another by a common push rod (2).
Port 1 is connected to compressed-air supply circuit 24 of the 4-circuit protection valve and is constantly supplied with compressed air.

Operating principle

If the driver wishes to open or close a door, he presses the corresponding push-button which sends an electrical control pulse to the solenoid valve. This causes the electromagnet (1) to force down the push rod (2) whose stem pushes its rounded end into the sliding surface of

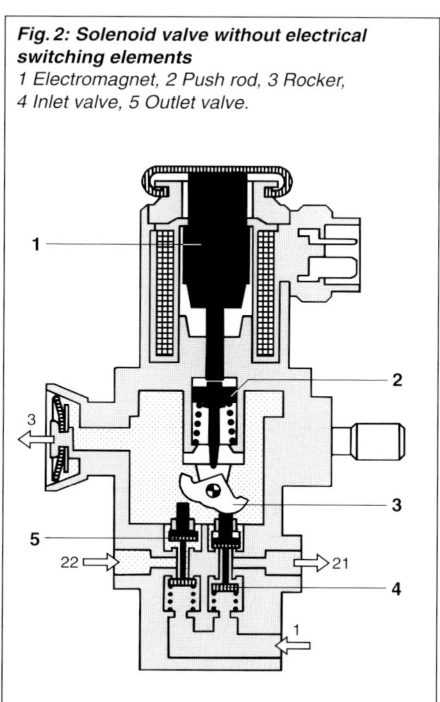

Fig. 2: Solenoid valve without electrical switching elements
1 Electromagnet, 2 Push rod, 3 Rocker, 4 Inlet valve, 5 Outlet valve.

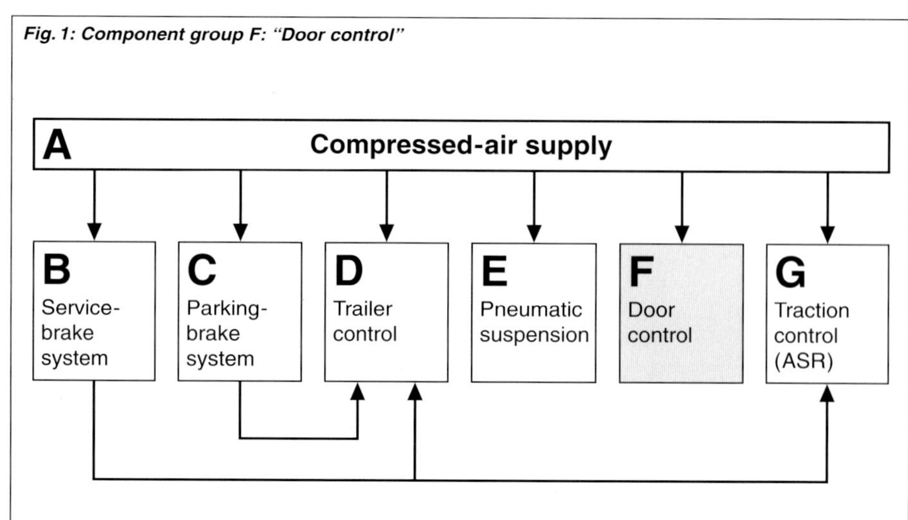

Fig. 1: Component group F: "Door control"

the rocker (3) and switches this to its other end position. This causes the inlet and outlet valves (4 and 5), which were previously open, to close, and those valves which were previously closed to open. In the working cylinder, the cylinder chamber which was pressurized is now depressurized and the depressurized chamber is pressurized. The air from the depressurized cylinder chamber escapes via port 3 of the solenoid valve.

Rotary actuator

Function
The rotary actuator controls operation of outside pivoting doors on passenger buses.
It opens and closes the door through a lower mounting joined to the vehicle's body.

Design
The rotary actuator is a double-acting cylinder capable of converting the lateral movement of its piston into rotary motion at the driveshaft.
The main components of the pneumatic rotary actuator (Fig. 3) are the:

– housing,
– driveshaft,
– anti-rotation element, and
– piston.

The rotary actuator also features two pneumatic fittings:
Connection 11 is used to close the door, while connection 12 is employed to open it. These connections control the supply and discharge of compressed air to the door-section drive. Integrated in the air fittings are throttle-type non-return valves; these valves make it possible to vary both the opening and closing speeds of the door.

Operation
With the door closed, compressed air is applied to connection 11 on the pneumatic rotary actuator, while connection 12 remains in the discharge mode. When compressed air is fed to connection 12 to open the door, the piston (7) is pushed upward. The pitch in the recirculating-ball path transfers the piston's linear stroke to the driveshaft. The anti-rotation element (6) in the bearing cage located between the piston and the housing prevents the piston from also rotating.

To close the door again, compressed air is directed to connection 11, while connection 12 reverts to the discharge mode. The piston (7) moves down and turns the driveshaft in reverse. The door shuts.

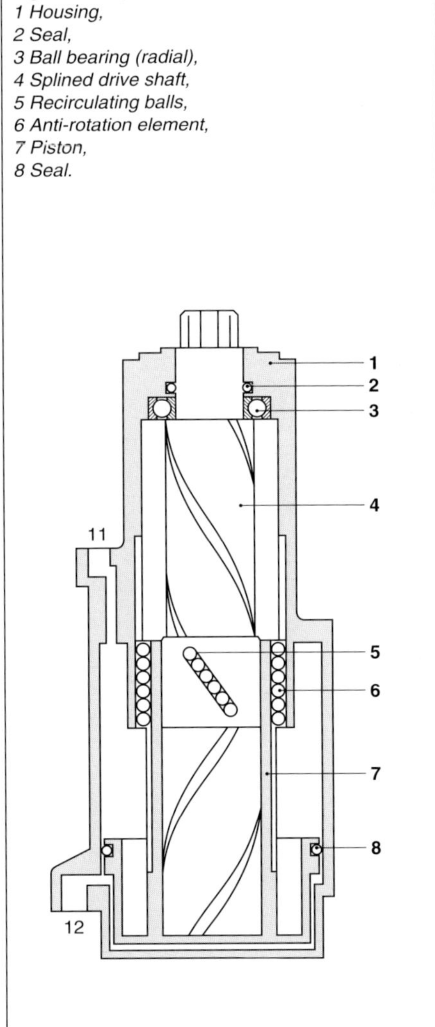

Fig. 3: Rotary actuator
1 Housing,
2 Seal,
3 Ball bearing (radial),
4 Splined drive shaft,
5 Recirculating balls,
6 Anti-rotation element,
7 Piston,
8 Seal.

Component Group G "ASR – Traction control"

Function

ASR is a traction-control system designed to inhibit wheelspin during acceleration, and provide optimal transfer of tractive forces to the road surface. ASR is an expanded version of ABS.

This traction-control system consists of closed-loop brake and engine control circuits. ASR is integrated within the ABS control unit, where it shares such ABS components as wheel-speed sensors and pressure-control valves. Supplementary components required for the ASR brake-control circuit include a 2-way directional control valve (shuttle valve) and a solenoid valve for each axle side. In the engine control circuit, a final-control element (such as a servo motor) is required to implement the required reductions in engine torque.

Operating modes

Brake control circuit

During initial acceleration on a "μ-split" road surface (characterized by relative variations in surface conditions at crown and shoulder of the road), the wheel on the surface with the lower friction coefficient can start to slip. The low friction coefficient at the slipping wheel leads to a reduction in tractive (accelerative) forces.

The brake controller at the affected wheel responds by generating torque that, through the differential, acts as drive torque at the stationary wheel. The brake control circuit uses this process to increase the effective tractive force available for moving the vehicle.

The solenoid valve applies braking force to the wheel on the low-friction surface via the pressure-control valve. This makes it possible for the wheel on the higher-friction surface to transfer drive torque to the road. The function of the brake control circuit is analogous to that of a limited-slip differential.

Engine control circuit

When the driver applies excessive throttle, the drive wheels will tend to spin on a homogeneous road surface affording only limited traction on both sides. The tractive force available to accelerate the vehicle is a function of the decreasing friction coefficient within the instable static-friction slippage range. Here the engine control circuit responds by reducing the engine torque to a suitable level in order to enhance traction and restore vehicle stability. Two types of electrical controller serve as the final-control element responsible for implementing these engine-torque reductions:
– an interface in the circuit controlling the electronic engine management system, or
– direct control via an electric ASR actuator (e.g., servomotor).

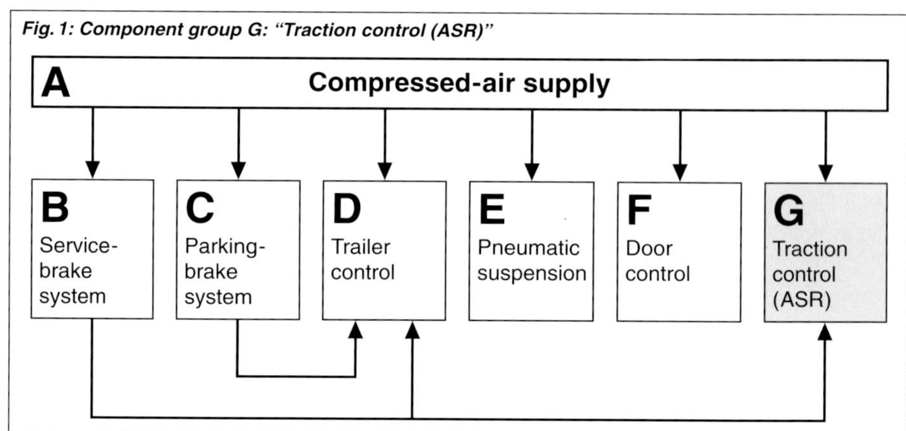
Fig. 1: Component group G: "Traction control (ASR)"

Interface

The ABS/ASR control unit receives signals indicating driver intention (consisting of factors such as accelerator-pedal setting and fuel-injection quantities) from the engine-management ECU. The ABS/ASR control unit uses this signal and other factors such as wheel slip in calculating the torque-reduction request that it transmits to the engine-management ECU, which responds by carrying out the request. Engine-management ECU's include EMS (Elektronische Motorleistungssteuerung or Electronic torque control) and EDC (Elektronische Dieselregelung, or Electronic diesel control).

Direct control within engine control circuit

The ABS/ASR control unit is connected directly to a servomotor. This motor, in turn, is mounted directly on the injection pump's stop lever. The ABS/ASR control unit responds to slippage monitored at the drive wheels by transmitting a control command to the servomotor, which reacts by immediately reducing the torque being generated by the vehicle's engine.

Components

Control unit (ECU)

The ABS/ASR control unit is a digital design featuring back-up (redundant) microprocessors. The unit's operation can be divided into four separate function groups:
Input stages, processor unit, output stages (output amplifiers) and power supply.
The control unit includes a C3 input connection to satisfy the legal requirements for road-speed limiters (Fig. 2).

ASR servomotor

The ABS/ASR control unit regulates the ASR servomotor directly. The servomotor is a DC unit with integral position feedback to ensure precise position control which remains independent of actuating forces at the injection pump, friction in the throttle linkage or any other interference factors. ASR intervention at the throttle linkage is restricted to reduction only, and inadvertant application of excess throttle is ruled out (Fig. 3).

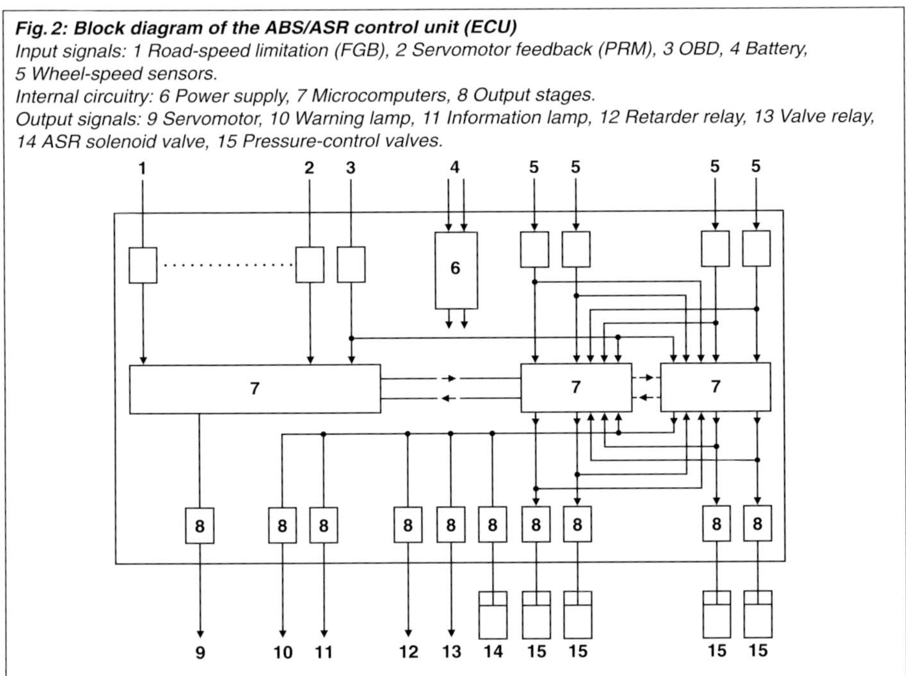

Fig. 2: Block diagram of the ABS/ASR control unit (ECU)
Input signals: 1 Road-speed limitation (FGB), 2 Servomotor feedback (PRM), 3 OBD, 4 Battery, 5 Wheel-speed sensors.
Internal circuitry: 6 Power supply, 7 Microcomputers, 8 Output stages.
Output signals: 9 Servomotor, 10 Warning lamp, 11 Information lamp, 12 Retarder relay, 13 Valve relay, 14 ASR solenoid valve, 15 Pressure-control valves.

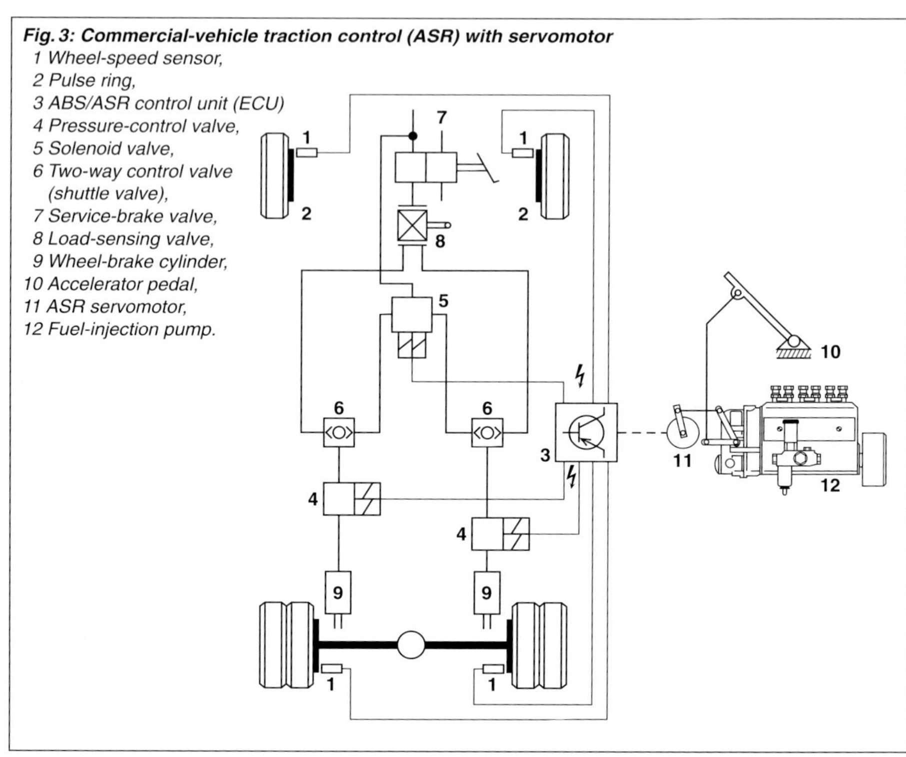

Fig. 3: Commercial-vehicle traction control (ASR) with servomotor
1 Wheel-speed sensor,
2 Pulse ring,
3 ABS/ASR control unit (ECU)
4 Pressure-control valve,
5 Solenoid valve,
6 Two-way control valve (shuttle valve),
7 Service-brake valve,
8 Load-sensing valve,
9 Wheel-brake cylinder,
10 Accelerator pedal,
11 ASR servomotor,
12 Fuel-injection pump.

Pressure-control valve

Each wheel is equipped with its own pressure-control valve, consisting of two solenoid-controlled diaphragm valves. Under normal braking, an unobstructed current of compressed air flows through the pressure-control valves on its way to the wheel-brake cylinders. The ABS/ASR responds to incipient slippage at one of the wheels by transmitting a signal to the corresponding pressure-control valve, which responds by increasing the pressure within the wheel-brake cylinder to brake the wheel (ASR). If a wheel starts to lock under braking, the signal to the pressure-control valve initiates a reduction in pressure within the wheel-brake cylinder (ABS).

Solenoid valve (Fig. 4)

The solenoid valve is actually a combination of two pilot-controlled 3/2 solenoid valves. Each of these applies compressed air to the wheel-brake cylinder of the drive wheel that is threatening to slip. These solenoid valves provide pilot con-

Fig. 4: Solenoid valve
1 Control piston, 2 Solenoid valve.
Connections:
(1) from supply reservoir,
(21) and (22) to the pressure-control valves,
(3) Atmosphere.

trol for two control pistons. When the solenoids are de-energized, supply 1 is blocked, while the cylinder connections (21 and 22) remain connected to atmosphere (3). ASR energizes the solenoid valve for the slipping wheel, which then relays supply pressure to the pressure-control valve.

Brake testing

Inspection and maintenance procedures

Because the condition of a vehicle's brake system exercises an immediate and decisive effect upon the safety of its passengers and cargo, brake-system maintenance and service assumes a high priority among vehicle maintenance procedures.

German law prescribes mandatory brake-system inspections at regular intervals. Inspection, maintenance and repairs can be performed by automotive dealerships as well as officially-approved and appointed service centers and brake-repair facilities (such as Bosch Service Centers). Official qualification requirements defined by the respective state government are in force for the service-center workshops and for their personnel.

In Germany, vehicle owners are responsible for and bear the cost of having their vehicles examined in an inspection (HU) at an officially approved inspection station (TÜV, DEKRA, etc.). The month in which the current approval expires and renewed inspection will be due is indicated by a sticker.

Buses with seating capacity for more than 8 persons, and heavy commercial vehicles with an approved gross vehicle weight in excess of 6 metric tons, are subject to the following supplementary requirements:

– Intermediate inspections (ZU) to ensure operating safety, and to check noise and exhaust emissions, and

– Special brake inspections (BSU) to ensure operational safety of the brake system.

Because the brake system is subject to natural wear (for instance, due to abrasion of the friction linings) the official inspections must be supplemented by regular maintenance.

This includes testing brake performance on roller-type brake analyzers, as well as performing operation checks and maintenance on the following components:
– Brake pads and linings,
– Brake discs (rotors), and
– Brake drums.

On hydraulic brake systems, inspection and maintenance procedures also include the following:
– Master cylinder,
– Wheel-brake cylinders,
– Brake hoses,
– Brake lines,
– Brake-fluid level, and
– Brake-fluid condition.

On compressed-air brake systems, the following components should also be inspected:
– Air compressor,
– Air tanks/reservoirs,
– Antifreeze unit,
– Valves and cylinders,
– Pressure regulator,
– Load-sensing valves,
– Couplings (gladhands), and
– Leakage check embracing the entire system.

Wheel brakes

Brake friction linings

Of all of the components in the brake system, it is the friction linings that are exposed to the highest rate of wear; these are the components that press against the discs (rotors) and drums to decelerate the vehicle. Correct maintenance is essential for ensuring brake-system safety.

<u>Wear inspection</u>
Assuming that the shoes or brake pads are correctly installed, the rate at which the friction linings abrade is determined by the lining composition (e.g., friction properties), by driving style, and by the vehicle loading.

On the disc brakes of many vehicles, the wheels must be removed before a reliable examination of pad wear can be performed. Attempts to evaluate pad wear with the wheels mounted can lead to false conclusions.

On most vehicles with drum brakes, both the wheels and the brake drums must be removed to check wear.

Some modern vehicles are equipped with sight holes in the backplate to allow visual inspections with the wheels still mounted, although wheels and drums must still be removed for comprehensive diagnosis.

<u>Adjustments</u>
A certain amount of clearance is required between the friction lining and the disk (rotor) or brake drum in order to ensure that the brakes do not continually "grind". As this clearance normally increases due to wear, the brake shoes require periodic adjustment (on vehicles not equipped with an automatic adjustment mechanism).

Disc brakes with integral parking brakes have automatic adjustment.

The basic disc brake adjusts itself automatically, the pistons adjusting automatically to maintain a constant clearance between pads and discs (rotors).

An increase in pedal travel indicates when adjustment is due on vehicles equipped with drum brakes without automatic adjuster.

It is essential that the manufacturer's instructions be followed when adjusting the brake shoes on drum brakes.

Regardless of design though, it always applies that the shoes on both sides of the vehicle must all be adjusted at the same time. If the vehicle is equipped with

drum brakes at both front and rear, then the procedure should always include adjustments at all four wheels.

The brakes should be adjusted when cold. The service (foot) brakes should be adjusted prior to the parking brakes.

<u>Friction-lining replacement</u>
The pads on disk brakes are due for replacement once the friction surface wears down to 2 mm.

If the vehicle is equipped with a pad-wear indicator (sensors in the pads), an indicator lamp in the instrument cluster will alert the vehicle operator when the pads reach a residual thickness of 3.5 mm, meaning that a pad change is imminent.

With drum brakes, the friction lining should not be allowed to wear beyond 1.5 mm on passenger cars, or 4 mm on heavy commercial vehicles. If the friction linings display varying wear patterns, if cracks or fissures are present, or if pieces have been broken off, then they should always be replaced.

When the friction linings are replaced, it is essential that the new units correspond to the manufacturer's specifications as contained in the vehicle's official homologation documentation (Germany: Allgemeine Betriebserlaubis, or ABE).

Important!
Installation of brake pads which do not carry official approval invalidates the vehicle's operating approval and insurance coverage.

Brake friction linings and discs (rotors) should never be replaced individually; replacement must always include all units on a given axle. Failure to observe this precaution could cause the vehicle to "pull" when being braked.

Every time the brake pads or shoes are replaced, the discs (rotors) and/or drums should be inspected for signs of the following:
– fissures,
– scoring,
– corrosion,
– wear, and
– variations in thickness.

Brake discs (rotors) and drums

Because the discs and drums are either castings or are manufactured from steel, they are less sensitive to wear than the friction linings. Despite this, they still require regular periodic maintenance.

It is important to examine the contact surfaces on the discs (rotors) and drums for wear, grooves, fissures and rust. These problems can be detected with a visual inspection.

Discs can also be examined for vertical and lateral runout. Runout, which should not exceed 0.2 mm at the disc's periphery, should be checked using a dial gauge. Discs with excessive lateral runout must be replaced.

When brake discs are machine-turned to remove grooves or recesses, it is important not to go beyond the minimum approved rotor thickness.

Brake drums are subject to runout and may also have hairline cracks. The runout stems from overheating, and one clue to its existence is brake-pedal pulsation; it can also be detected on the brake analyzer. Depending upon the amount of wear or damage, it may be possible to turn the drum on a lathe, whereby the machining specifications for the particular vehicle should always be observed. If the type or extent of the damage are such as to prohibit turning, then the brake drums must under all circumstances be replaced.

Brake-drum replacement must always include all drums on both sides of the axle in order to ensure balanced braking response.

Hydraulic brake systems

The hydraulic brake system includes the master cylinder, wheel-brake cylinders, brake lines and hoses, all of which require maintenance. Various vehicle types are also fitted with additional equipment such as brake boosters, brake-force distribution and proportioning devices, etc. – these are usually maintenance-free. The following components should be inspected and serviced at regular periodic intervals:

Master cylinder

The primary wear components in the master cylinder are the seals between the pistons and the cylinder walls, which are composed of a special rubber compound. Corrosion of the kind encountered when the brake fluid absorbs moisture causes roughening of the cylinder walls, with the attendant negative effects on the seals, which then wear and start to leak.

Depending upon the extent of the damage, it may become impossible to maintain brake pressure, or to even generate it in extreme cases. The response at the pedal indicates whether the primary or intermediate seal is damaged.

Wheel-brake cylinders

As in the master cylinder, in the wheel-brake cylinder it is the seals that are exposed to wear. They can start to leak, resulting in corrosion on the cylinder walls. Leakage can also occur on the wheel-brake cylinder, e.g., at the sealing caps. Brake fluid can then emerge, finding its way to the friction linings and leading to a deterioration of braking efficiency.

The following procedures can be employed to check the condition of the seals:

Low-pressure test

A pressure gauge is connected to the wheel-brake cylinder. Next, a special device is attached to the pedal and used to apply a pressure of 2...5 bar. The pressure should remain constant with no drop for a period of 5 minutes.

High-pressure test

A pressure of 80...100 bar is applied. At the end of ten minutes the pressure should not have fallen by more than 10%.

Latent-pressure test

The pressure device is removed from the pedal. The system falls to its latent pressure (if present; only on cylinders with cup seals) of 0.4...1.7 bar. After 5 minutes, the pressure should be at least 0.4 bar.

Brake hoses and lines

The brake hoses and lines are basically wear and maintenance-free. At the same time, they are exposed to corrosive influences in the form of moisture and road salt, and can be damaged by stones and gravel, etc.

Due to these high levels of physical stress, brake lines and hoses should be examined regularly. On the brake lines, the emphasis is on corrosion, while the hoses should be inspected for abrasion and cracks. All connections should be inspected for leaks.

Brake-fluid level and condition

The brake-fluid level is checked at the reservoir; the fluid should extend up to between the MIN and MAX graduations. This check can provide a warning of leaks in the system. If the fluid has fallen to the MIN graduation or below, the brake system should be examined for leaks. On many vehicles an indicator lamp warns the driver of low brake fluid.

Diffusion at the hoses can allow the brake fluid to absorb atmospheric moisture; the brake fluid should thus be replaced every one to two years.

This regular brake-fluid replacement plays an essential role in ensuring that the brake system continues to operate safely.

Compressed-air brake systems

The compressed-air (or pneumatic) brake system includes the following components: Compressor, air tank/reservoir, antifreeze pump, valves, cylinders, pressure and brake-force regulators. This type of system is employed exclusively in commercial vehicles. The compressed-air brake system requires daily maintenance in the form of an inspection prior to leaving the lot. This inspection encompasses the following individual components:

Compressor
If the compressor is not equipped with an automatic (pressurized line) oiler, then the oil level should be checked to see whether lubricant must be added by hand. The oil should reach up to between the two graduations on the dipstick.
The drive-belt tension should also be checked.

Air tank(s)
If the air tanks/reservoirs are equipped with manual drain valves, then these should be opened so that accumulated condensate can escape.

Antifreeze device
If it is possible that extreme cold will be encountered, then the antifreeze device should be set for operation (not required on automatic units).
The level of antifreeze in the reservoir should also be checked.

Valves, cylinders
The operation of the valves and cylinders should be checked.
As an example, the trailer-control valve should transmit 0.8...1.5 bar to the trailer-brake line at a brake-valve pressure of 1 bar.

Pressure regulator
Check the operation of the pressure regulator. An audible air "blow-off" indicates the point at which cycle (deactivation) pressure is reached.

The activation pressure has been reached once the needle on the system-pressure gauge starts to climb. With the engine running, the pressure regulator should not cycle at intervals of less than approximately 2 minutes. The difference between activation and deactivation pressures should be 0.5...1.1 bar.
The low-pressure warning device should be checked while the air tanks are being filled.

Brake-force regulator
Adjust the lever on the trailer's brake-force regulator (load-sensing valve) in accordance with trailer load. This procedure is not required on ALB-equipped vehicles – here it is only necessary to check the control valve's shaft for ease of movement.

Coupling heads (gladhands)
The covers on the coupling heads should always be closed when no trailer is connected.
Before a trailer is connected, the seals on the coupling heads should be inspected to ensure that they are in good condition before the lines are connected.

Leaks
With an initial application pressure of approximately 3 bar, the drop in supply pressure after 3 minutes should not exceed 5%.

Brake test stands
(dynamic brake analyzers)

The vital significance of the brake systems for vehicle safety makes it imperative that they are regularly tested. When these tests are carried out in accordance with §29 StVZO (FMVSS/CUR), or during inspections and repair in the workshop or at the brake service station, the vehicle is usually driven onto a roller-type dynamic brake analyzer (Fig. 1). In Germany, the test stands for inspection and testing as per §29 StVZO (FMVSS/CUR) must comply with the "Regulations governing the use, design, and testing of brake test stands" as issued by the Federal Transport Minister.

Design

The main components of the roller-type brake test stand are two mutually independent roller sets for the vehicle's left and right sides respectively. The vehicle is driven onto the test stand so that the wheels of the axle under test rest on the rollers (Fig. 1).

A stable frame supports the roller sets which are in the form of a drive roller and a secondary roller in a parallel layout. A chain is used to provide the positive dynamic connection between the two rollers. The drive roller is powered by an AC motor through a step-up ratio gear set, the drive unit itself being suspended on an extension of the drive roller's shaft. Pressure exerted against a torque lever flanged onto the gear-drive unit is transferred through the load sensor, with the frame providing positive support for the entire assembly. The braking force F_{Br} is actally measured by monitoring the reaction torque M_R. The electric motors set the rollers in motion and then maintain a constant rotational speed against the considerable opposing forces that are generated when the vehicle's brakes are applied. The suspended drive unit with torque lever transmits the braking forces to the load-sensing device which can be in the form of an aneroid unit, incorporated within the hydraulic system, which acts directly upon a gauge. The gauge's scale is calibrated in newtons and provides an analog display of the braking force.

Electrical load-sensing systems can employ flexural sensors with a wire strain gauge, or they can be designed to use an inductive short-circuit ring sensor operating in conjunction with a linear-motion spring strut.

The system's computer employs digital technology in the evaluation of all the data registered during braking-force testing, for instance fluctuations or differences in braking force. Once processed, the information is clearly displayed in either analog or digital form depending upon the system. A printer can be connected to provide a hard-copy test protocol.

Operation

The roller-set drive motors are switched on either by remote control or by the test stand's automatic On-Off switch. Visible indication of such an automatic on-off facility is provided by the sensor rollers located between the main test rollers on each roller set. When driven onto the test stand, the vehicle forces down the sensor rollers and switches on the test stand. Conversely, when it is driven off again the vehicle releases the sensor rollers and automatically switches off the test stand again. As soon as the applied braking force starts to exceed the available traction between tires and test rollers, the wheel in question will respond by starting to slip, and then it will lock. Tire slip, however, makes it impossible to perform useful measurements of braking force.

Under slip conditions, it is the slip resistance between tire and roller (as a function of wheel load) that is measured and this data is of no use at all. Such false readings, and the possibility of tire damage, are avoided by the "automatic wheel-slip shut-off device". This monitors the slip via the sensor-roller speed and switches off the test stand when a set maximum slip is exceeded.

At the instant before the test stand is switched off, the braking-force display shows the maximum braking force being generated by the brakes. On analog displays, an indicator lock is used to maintain the reading when the test stand is switched off, while on digital displays the electronic memory stores the data for read-out. Both arrangements ensure that the final display remains visible long enough for the operator to record it.

In addition to the braking forces, the vehicle's weight and/or axle weight can be entered by remote control, or measured on the test stand. This data can then be used for calculating the effective retardation.

The automatic sequences used on the test stand ensure an extremely efficient brake-testing procedure. The operator can carry out the complete testing of the front and rear-axle brakes without having to leave the vehicle.

Vehicles with permanent all-wheel drive and variable torque distribution are tested on special stands which are designed to prevent the forces generated at the test axle from being transferred to the axle which is not turning.

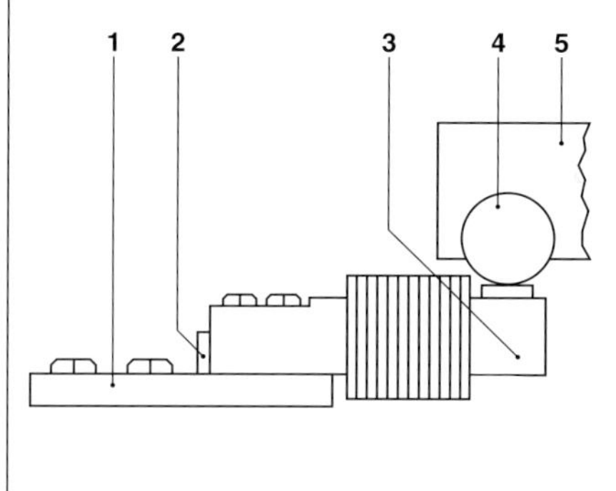

Fig. 2: Measuring sensor in the brake test stand
1 Adjustment plate,
2 Alignment pin,
3 Flexural sensor with wire strain gauge,
4 Thrust block,
5 Torque lever.

Fig. 1: Measuring the braking force F_{Br}
By measuring the reaction torque M_R.
1 Vehicle tire, 2 Roller set with spacing a, 3 Motor with gear set,
4 Torque lever with length l, 5 Measuring sensor, 6 Display.

Index